The Earth

Its Birth and Growth

Second Edition

Recent environmental problems and natural disasters have given cause for increasing concern over the future habitability of our planet. It is becoming increasingly apparent that a clear understanding of the Earth's past evolution can provide the key to its possible future development. *The Earth: Its Birth and Growth* explores the evolution of the Earth over 4.6 billion years using basic reasoning and simple illustrations to help explain the underlying physical and chemical principles and major processes involved.

Now fully updated and revised, this rigorous but accessible second edition includes three completely new chapters and additional illustrations. It incorporates recent exciting developments in isotope geology, placing results from these advances within a wider framework of Earth evolution and plate tectonics. Some background in physics and chemistry is assumed, but basic theories and Earth evolution processes are explained concisely in self-contained sections. The book also illustrates specific topics with short accounts of the work of eminent scientists at different stages of discovery in the field. Key research papers and review articles are fully referenced in each chapter to enable readers to explore further.

This book is ideal as a supplementary text for undergraduate and graduate students in isotope geochemistry, geodynamics, plate tectonics, and planetary science. It also provides an enjoyable overview of the Earth's evolution for professional scientists and general readers.

MINORU OZIMA is an Emeritus Professor in the Earth and Planetary Science Department at the University of Tokyo. He was awarded the prestigious V.M. Goldschmidt Medal in 2010, recognizing his major achievements in geochemistry and cosmochemistry. Professor Ozima was among the first to focus attention on the information contained in

noble gas isotopes in application to the formation and evolution of the planets. He is a leading figure in this field, having contributed significantly to the establishment and development of the geochemistry and cosmochemistry of noble gases. He has published several books and is a Fellow of the American Geophysical Union, the Meteoritical Society, the European Association of Geochemistry and the Geochemical Society.

JUN KORENAGA is a Professor of geophysics at Yale University where he studies the evolution and dynamics of the Earth with a variety of theoretical and observational techniques. Professor Korenaga is particularly known for his new theory of the Earth's thermal history and, in recognition of his contributions, was awarded the James B. Macelwane Medal in 2006 from the American Geophysical Union. His current research spans mantle and core dynamics, theoretical geochemistry, and marine geophysics, and he is also extending his work to cover other Earth-like planets within and outside the Solar System.

QING-ZHU YIN is a Professor in the Department of Geology at the University of California, Davis. Having received his Ph.D. with highest distinction from the Johannes Gutenberg University and Max-Planck-Institute for Chemistry in Mainz, Germany, he expanded his research experience at the Department of Earth and Planetary Sciences at Harvard University. His research interests now range from the use of isotopes to study the formation of the Solar System, to isotope and trace element geochemistry with applications to crust mantle evolution. Professor Yin is the author or co-author of over 60 research articles, and is a member of the Geochemical Society, the American Geophysical Union, and the Meteoritical Society.

The Earth

Its Birth and Growth

Second Edition

MINORU OZIMA
Department of Earth and Planetary Science, University of Tokyo

JUN KORENAGA
Department of Geology and Geophysics, Yale University

QING-ZHU YIN
Department of Geology, University of California, Davis

CAMBRIDGE UNIVERSITY PRESS
Cambridge, New York, Melbourne, Madrid, Cape Town,
Singapore, São Paulo, Delhi, Mexico City

Cambridge University Press
The Edinburgh Building, Cambridge CB2 8RU, UK

Published in the United States of America by Cambridge University Press,
New York

www.cambridge.org
Information on this title: www.cambridge.org/9780521760256

First published 2012

Printed in the United Kingdom at the University Press, Cambridge

A catalog record for this publication is available from the British Library

ISBN 978-0-521-76025-6 Hardback
ISBN 978-1-107-60076-8 Paperback

Contents

Preface to the second edition

A few years ago, Professor David Hilton of the University of California, San Diego mentioned to me that he was still using my book, *The Earth: Its Birth and Growth*, as suggested reading in his class. The book was published in 1979 by Cambridge University Press. Amazed by its longevity, I became curious about how this seemingly plain small book could have survived in the recent swarm of the media world, in which there are a flood of books on astounding findings in Earth and planetary sciences with colorful pictures and illustrations. I read the book once again, and I was convinced that it was worth revising it by incorporating recent developments.

The new edition has therefore attempted to keep the original style of the first edition: that is, to maintain readability without sacrificing scientific rigor. The concise style of the book is important so that readers can see the big picture without being drowned by a formidable amount of information. Obviously many of the materials in the first edition needed to be updated. Also, given recent developments, I wanted to emphasize in the new edition the importance of integrating a vast range of geophysical and geochemical data to develop a coherent view of Earth's evolution.

To update the book as planned above, I first asked Qing-zhu Yin for help, but because of his hectic schedule, he suggested asking Jun Korenaga to join the project. When Jun received the Macelwane medal from the American Geophysical Union, he mentioned that his interest in studying the history of the Earth originated in attending my unorthodox geophysics course taught at the University of Tokyo many years ago, so asking Jun for help seemed quite suitable. Preparation of the new edition benefited greatly from his enthusiasm for the project, both in speed and quality. Qing-zhu Yin examined a draft at various stages to improve its accuracy and clarity.

Although the book is primarily aimed at general readers, we did not hesitate to include some of our own ideas such as are seen in Chapter 10, since a seriously curious audience, whether science-minded or not, is keen to learn a logical way of thinking rather than to read a mere description of facts. Some of the bold ideas should be inspiring to experts as well. We believe that this small book should answer many basic questions by general readers on Earth's evolution, such as how and when the Earth formed, with as much rigor and brevity as is allowed within the scope of the book, while also being useful to those who specialize in this discipline.

Minoru Ozima

Preface to the first edition

It is thought that the Earth was born as a planet about 4500 million years ago. Throughout the long years since then it has continually evolved, and has undergone a transformation into its present form.

Tracing the evolution of the Earth is a central topic in Earth science, and has been dealt with by many writers. However, most previous histories of the Earth have been concerned with the past 600 million years, since fossils have been found in abundance from this period, and only touch very briefly on the Precambrian period, which is equivalent to roughly seven-eighths of the Earth's history. But those basic qualities of the Earth with which we are so well acquainted – the magnetic field, the layered structure of the core and mantle, the atmosphere and oceans, were all formed in the very early stages of the Earth's history.

Until 1950 virtually nothing was known about the early form of the Earth, but with the appearance of isotope geochemistry using radiogenic elements, it is gradually being brought to light. This book describes the birth of the Earth and its growth, outlining the problems which are now being solved rapidly through isotope geochemistry. It is an attempt to sketch the evolution of the Earth over 4500 million years.

In writing this book, I received much helpful advice from Drs Sadao Matsuo, Kiyoshi Nakazawa, Kenji Notsu, and Naoki Onuma. I would also like to express my deep gratitude to Mr Toshio Ogawa and Ms Yuko Natori of Iwanami Shoten Publishers, who spared no efforts in producing this book.

Minoru Ozima
December 1979

Preface to the English edition

To cover the 4500 million years of the history of the Earth in one book is certainly a formidable task. As my particular field lies in isotope geochronology and rock magnetism, which are the most effective means of clarifying the Earth's evolutionary history, I have been able in this book to present my own view of the Earth's evolution mainly on the basis of results obtained by these two approaches.

In preparing the English edition, I have made a few changes following comments by my colleagues on the original Japanese edition. I have now realised that to prepare the English edition involved far more than mere translation. I have had to admit that the Japanese language is more suited to literature than it is to being a scientific medium. So for Mrs Judy Wakabayashi the task was to convert a language suited to the heart into a language suited to the mind. And as far as the English edition is concerned, I feel that she is almost entitled to be a co-author, and I would like to express my very deep appreciation of her work and for all the "blood, sweat and tears" which she has endured during the past six months.

Minoru Ozima
Tokyo, September 1980

1   Heat from within: energy supporting the dynamic Earth

THE ENERGY HIDDEN IN ROCKS

Granite is one of the most common and well-recognized rocks to occur at the surface of the Earth. Let us suppose that we put a fragment of granite in a small container, which is then completely sealed. We will assume that the container is made of an ideal thermally insulated material, and that heat can neither escape from within nor enter from outside. What changes will occur in the granite inside the box?

If the contents of the box were examined after one or two years, probably no changes at all would be observed. However, if it were examined after the passage of several hundreds or thousands of years, a careful observer would no doubt realize that the temperature within the container was rising very slightly. After a few hundred thousand years, this rise in temperature would be apparent to any observer. If the calculation described later is carried out, it is clear that the granite in the sealed container would melt completely after several tens of millions of years owing to the rise in temperature.

Where is this energy hidden in the seemingly commonplace granite? Granite contains minute quantities of uranium and thorium, and these radioactive elements are the source of the energy. In general, granite contains several parts per million (ppm) of uranium and about 10 ppm of thorium. The nuclei of these elements undergo radioactive decay, and over their very long half-lives, which span 700 million to 14 billion years, gradually change into isotopes of lead, which have stable nuclei. When the uranium and thorium nuclei decay, alpha-particles, electrons, and other particles are emitted at high speed. These particles are able to move through the mineral's crystals for just a few micrometers (millionths of a meter, often called microns), colliding with the surrounding atoms and coming to a halt once all of their kinetic energy has been transferred as heat to the mineral crystals. This heat is what melts the granite within the sealed box.

In addition to uranium and thorium, one isotope of potassium, potassium-40, is another important radioactive nuclide in rocks. It has a half-life of 1250 million years, and undergoes natural radioactive decay into argon-40 and calcium-40, by electron capture and electron loss (beta decay) respectively. Potassium-40 accounts for only about 0.01 percent of all potassium, but since granite contains quite a lot (several percent) of potassium, potassium-40 is also present in significant quantities.

If the energy released by such radioactive elements is totalled, approximately two-hundredths of a joule of heat is generated per year per kilogram of ordinary granite. The existence of radioactive elements is not limited to granite. Basalt also contains uranium, thorium, and potassium, but in amounts about one order of magnitude less than those in granite. Several hundredths of the amount of radioactive elements in ordinary granite are also contained in peridotite, which is the main component of Earth's mantle, though it is not usually seen on the surface. The amount of radiogenic heating within the Earth is extremely slight, being on average about a ten-thousandth of a joule per year per kilogram. In the long run, however, an enormous amount of heat has been produced throughout Earth's history. Repeated volcanic eruptions, earthquakes, mountain-building, and tectonic movements – these events that characterize the dynamic Earth would not happen today without the energy released by the nuclear disintegration of these radioactive elements, which keeps the Earth's interior burning hot.

THE AMOUNT OF RADIOACTIVE ELEMENTS IN ROCKS
Uranium and thorium are part of the third group in the periodic table, and they are also very heavy elements. In normal conditions, both of these elements become ionic with a valence of +4. The ionic radius for uranium is 0.85 Å, and that for thorium is 0.94 Å. The electronic charges of uranium and thorium are equal, and they are almost the same size, so chemically they behave in a very similar way. Consequently, in rocks and

Table 1.1 *Amounts of uranium, thorium, and potassium contained in rocks (typical values)*

	Uranium (ppm)	Thorium (ppm)	Potassium (percent)
Granite	3.0	12.0	3
Basalt	1.0	4.0	0.5
Peridotite	0.001	0.004	0.04
Sandstone	0.45	1.7	1.4
Deep ocean-floor sediment	1.3	7	2.5

ores in which uranium is concentrated, thorium is also concentrated. Even if the type of rock or ore differs, usually the ratio of uranium and thorium is almost the same (Table 1.1). But in an environment extremely susceptible to oxidation, uranium sometimes becomes ionic with a valence of +6. This uranium forms an ion (UO_2^{2+}) which is water-soluble; it is gradually washed from the rock and thus the uranium is depleted, so in some cases the ratio of uranium to thorium changes greatly.

Owing to their large ionic valence, it is difficult for uranium and thorium to be introduced into mineral crystals. When magma is generated by partial melting of the mantle material, therefore, almost all of the uranium and thorium are expelled from the original mineral crystal to enter the magma. The solidification of such magma near the Earth's surface forms the crust, and this is why uranium and thorium are much more concentrated in crustal rocks such as granite and basalt than in mantle rocks such as peridotite. Potassium is also incompatible with mineral crystals because of its large ionic radius (1.33 Å) and so is more concentrated in crustal rocks. A summary of the abundance of these three elements is given in Table 1.1.

The solid Earth is divided into three basic layers. The part closest to the surface is the crust, which accounts for less than 0.4 percent of

Earth's total mass. Below this is the mantle, which makes up approximately two-thirds of Earth's total mass, and the remainder is the central part, called the core. The average chemical composition of the crust is one in which granite and basalt are mixed in a ratio of roughly one to three. The chemical composition of the mantle is believed to be similar to that of peridotite. If that is the case, it is clear from Table 1.1 that the average amount of potassium contained in the crust is about 1 percent, and in the mantle about 0.04 percent. Likewise, there is about 2 ppm (parts per million) of uranium in the crust, but this falls to a few ppb (parts per billion) in the mantle.

RADIOACTIVE ELEMENTS IN THE WHOLE EARTH

We have already said that rocks and Earth are heated from within by the energy released when uranium, thorium, and other radioactive elements naturally undergo nuclear disintegration. We also said that if this energy is sealed up in the rock without escaping outside at all, the heat generated is sufficient to melt granite completely within several tens of millions of years. What would happen in the case of the whole Earth? In considering this question, it is first necessary to estimate the quantities of radioactive elements, such as uranium and thorium, in the whole Earth.

Although Earth's crust accounts for an extremely small part of the whole Earth in terms of mass, it is mainly composed of granite and basalt, in which radioactive elements are highly concentrated. The mantle also contains radioactive elements, though at a much lower level than the crust. The main component of the core is believed to be an alloy of iron and nickel. Though we do not have a direct sample from the core, we can still estimate its gross properties by studying how seismic waves propagate through the Earth and by considering the Earth's density as well as the similarity of its composition to that of meteorites. Elements like uranium and thorium hardly dissolve at all in the iron–nickel alloys found in meteorites. So the amount of radioactive elements in the whole Earth – that is, Earth's internal heat source – can be approximately estimated from the amount present in the crust and

Table 1.2 *Heat generation of radioactive isotopes (in unit of joules per year per kilogram)*

Uranium-238	2955
Uranium-235	17 944
Uranium as a whole*	3063
Potassium-40	880
Potassium as a whole[†]	0.112

* At present uranium-238 and uranium-235 form uranium in the ratio of 137.88 : 1.
[†] At present potassium-40 accounts for 0.0117 percent of total potassium.

mantle.[1] From Table 1.1 it is calculated that there is about 0.01 ppm each of uranium and thorium; that is, each gram of Earth as a whole contains approximately a hundred-millionth of a gram of uranium and thorium, respectively.

Table 1.2 shows the energy released when each of these radioactive elements undergoes natural nuclear disintegration. The uranium in nature consists of two isotopes, uranium-238 and uranium-235. The ratio between these two uranium isotopes ($^{238}U/^{235}U$) is surprisingly uniform, having a value of 137.88. Actually, all the isotopic ratios of elements in nature have fairly uniform values (with the exception of isotopes added by the process of radioactive decay). This is an extremely important constraint when considering the origin of the Earth and the Solar System, but we will leave this to be taken up in Chapter 2, and now proceed to discuss heat generation in the Earth's interior.

From the data shown in Table 1.2, the amount of heat generation in Earth's material is calculated to be approximately one ten-thousandth of a joule per year per kilogram. This is the present value, and naturally it increases as one goes back in time because radioactive elements were more abundant in the past. The total energy released through nuclear disintegration over the Earth's history would be more than several hundred kilojoules per kilogram of Earth material. The specific heat of normal rock is about one kilojoule per kilogram per degree, so this amount of nuclear energy could potentially raise the temperature of rock by more than a thousand degrees: that is, up to a point where it starts to melt.

HEAT GENERATION IN THE EARTH'S INTERIOR

With the exception of part of the core (from seismic observations, the outer part of the core, which accounts for about half the core radius, is known to be in a liquid state), the Earth is still a solid despite this enormous internal heat production. Why? This is of course because a considerable amount of energy has escaped from the Earth into space. Just as a cup of hot coffee gradually cools by releasing heat into the surrounding air, the Earth cools down by transferring heat from its hot interior to cold space. The internal heat production by radioactive elements slows this cooling, or if the heat production is greater than the heat release, the Earth can heat up. The balance between internal heat generation and surface heat loss thus determines the thermal fate of a planetary object.

The size of a planetary object places a first-order control on this balance. The loss of heat from the surface is proportional to the square of the radius of an object, while the amount of heat generated within an object is proportional to the cube of the radius. Therefore, the ratio of heat loss to heat generation is inversely proportional to the radius; the larger an object, the more heat can be retained within the object. For example, let us compare the Moon and Earth. Even if they contained roughly the same amount (per unit mass) of radioactive elements, the Moon cooled more rapidly than the Earth because it is much smaller. This is why almost no mountain-building or tectonic movements as seen on Earth occur on the Moon. As another example, consider meteorites found on the Earth, which are thought to be fragments of tiny proto-planets called "planetesimals". On these meteorite parent bodies, with average sizes far smaller than the Moon (they are believed to have radii of several to several hundred kilometers), the cooling rate should have been even faster. Before falling down to Earth, therefore, these meteorites have drifted for a long time through space as "dead bodies" that stopped "breathing" at the time of their birth. This is why meteorites are called the fossils of the early Solar System.

For a relatively large planetary object such as the Earth, the cooling history would not be so simply dominated by surface heat loss, and the balance between heat generation and heat loss can be a delicate one.[2] While the amount of heat generation is determined from the Earth's chemical composition, the rate of heat loss from the surface can be measured by geophysical techniques. Doctors diagnose the human body by means of tapping and body temperature. Using this analogy, analysis of the Earth's internal structure by use of seismic waves is equivalent to the doctor's tapping, with the earthquakes as the sources of signal, and measurement of the amount of heat escaping is similar to measuring body temperature. These measurements are an essential geophysical method in "diagnosing" the Earth.

The amount of heat escaping from the Earth can be estimated by measuring changes in temperature with descent through a mine shaft. As one descends through the mine levels deep underground, one can feel that the temperature gradually rises. The temperature increase varies from mine to mine, but usually the temperature has risen by about 10 to 20 °C at a depth of about 1 km. The amount of heat escaping from the interior to the surface can be calculated by multiplying the thermal conductivity of rock by the temperature difference between upper and lower levels per unit length as one proceeds downwards. Since the thermal conductivity of rock when actually measured in a laboratory is about 3 joules per meter per second per degree, the amount of heat escaping from the Earth is $3 \times 20/1000 = 0.06$ joules per square meter per second. That is, six-hundredths of a joule of heat escapes to the surface of the Earth per square meter every second.

The amount of heat escaping from the Earth's surface can be measured both for the land surface and for the ocean floor. The principle behind this measurement is the same as for measuring on land. A thermometer is inserted into the mud on the ocean floor, and the

temperature difference between depths of about 1–2 m apart is measured. Here the amount of heat escaping is calculated by multiplying
the thermal conductivity of the mud on the ocean floor by the temperature difference. Unlike continents, the temperature of the sea floor is
unaffected by seasonal variations because the heat capacity of the
ocean water keeps the sea floor at a relatively even temperature, so
we do not have to dig deeply to obtain stable temperature measurements. A vast quantity of data thus has been gathered so far from ocean
basins. The average amount of heat escaping on the ocean floor is
approximately one-tenth of a joule per square meter per second,
which is not much different from the average value on land.

From these measurements, the total heat loss from the entire
surface of the Earth is estimated to be about 46 trillion joules per
second.[1] As mentioned before, the average heat production of the
Earth is approximately a ten-thousandth of a joule per year per
kilogram, which is equivalent to about 20 trillion joules per second for
the entire Earth. Nearly half of the surface heat loss is thus compensated
by internal nuclear energy at present.[3]

COMPARISON WITH SOLAR ENERGY

In closing this chapter, let us now compare the internal heat with solar
energy, which is another important energy source for the Earth system
as a whole. The solar energy received at Earth, per unit time per unit
area that is perpendicular to the Sun's rays, is called the "solar constant", and is approximately 1.3 kilojoules per second per square
meter. The amount of solar energy that the whole surface of the
Earth receives, calculated by multiplying this constant by the cross-
section of the Earth, is approximately one joule per year per kilogram if
distributed within the entire Earth. This is about ten thousand times
greater than the heat generated by radioactive elements. However,
most of the solar energy is reflected back into space again or used up
to drive circulations in the oceans and atmosphere, and has almost no
effect on temperature within the Earth. So the energy released from
radioactive elements through nuclear disintegration is effectively the

sole energy source for the dynamic history of the Earth, which we will try to unfold in this book.

NOTES

1. The possibility that potassium may go into the core has been repeatedly discussed,[2–5] but such behavior by potassium would be accompanied by other elements of similar chemical nature, which is not observed.[6]
2. The latest estimate on the thermal history of Earth suggests that the planet was warming up instead of cooling down prior to about 3.5 billion years ago.[7]
3. The Earth's thermal budget is actually a controversial topic. Geophysical theories developed in the early 1980s [8,9] suggest that Earth's heat flux is predominantly supported by radiogenic heating, much more than indicated by the compositional models of Earth, and this view of a highly radioactive Earth is often considered as a standard model among geophysicists.[10] For a comprehensive review on this issue, see [11].

2   At the time of the Earth's birth

The colorful drama of Earth's evolution begins with the formation of the Solar System. The Earth is, after all, merely one of several planets orbiting around the Sun, and the story of how the Earth formed cannot be told without describing how the entire Solar System was formed. Studying the formation of the Solar System in turn enables us to better understand the processes leading to the birth and death of stars, and the mechanisms by which all of the chemical elements in the universe were formed.

Thanks to nuclear physics, we now know how the chemical elements in the universe were made. Except for hydrogen, helium, and some lithium, which were created shortly after the Big Bang, all of the chemical elements were made by stars.[1] Stars explode to end their life and eject the old and newly created elements into interstellar space. When a cloud of interstellar materials has high enough density, it will collapse under its own gravity to form a new star as well as a dusty disk around it, from which planets will emerge. A whole new cycle of chemical element synthesis (called nucleosynthesis) starts again within the central star, until the star consumes all the nuclear fuel and ends its life cycle explosively. In doing so, the universe becomes chemically more and more enriched with heavy elements.

Deciphering the details of nucleosynthesis in the universe, in particular how all of those elements that constitute the Earth were actually formed, is quite difficult, but we have some clues from isotopes. As mentioned in Chapter 1, the isotopic ratio of uranium-238 and uranium-235 has a present value of 137.88 : 1. What does this tell us? Based on nuclear physics, we can calculate the ratio of ^{238}U to ^{235}U in an exploding supernova, and the difference between the theoretical ratio and the observed ratio can be translated into the mean age of nucleosyntheses in our galaxy.[2] Similar calculations can be carried out for other isotopes; for example, the ratio of thorium-232 and uranium-238 can also be used. This type of calculation, however, tells us only a mean age. During the long half-lives of these uranium and thorium isotopes (several billion years and longer), many stars must have exploded and contributed to the interstellar medium from which our Solar System was born.

More stringent information can be obtained from a different kind of radioactive isotopes known as "extinct" isotopes. Iodine-129 (^{129}I), for example, has a half-life of only about 16 million years. That is, the abundance of ^{129}I is halved every 16 million years, so the abundance will decrease to about one-thousandth of the initial abundance after 160 million years, and ^{129}I will become practically extinct. The decay of iodine-129 produces xenon-129 (^{129}Xe), which is a stable isotope. The Earth and meteorites are found to contain radiogenic ^{129}Xe (which is denoted as ^{129}Xe* to distinguish it from non-radiogenic ^{129}Xe; we can distinguish between the radiogenic and non-radiogenic components based on the systematics of xenon isotopes), and because ^{129}Xe* can be produced only by the decay of ^{129}I, the existence of ^{129}Xe* means that the Earth and meteorites must have had a certain amount of ^{129}I when they were formed; that is, the Solar System must have formed shortly after the nucleosynthesis of ^{129}I, within 100 million years or so. Conversely, this means that the nucleosynthesis of ^{129}I took place no more than about 100 million years before the formation of the Earth and meteorites. This fact was first proved in the 1960s by J. H. Reynolds and his colleagues at the University of California and published in a classic paper titled "Xenology", announcing the arrival of a new discipline in cosmochemistry.[1]

That ^{129}I was produced roughly 100 million years before the birth of the Solar System, however, does not mean that all the other elements were also produced at that time. Rather, it is thought that the opposite occurred; the uranium and thorium isotope ratios indicate, as mentioned earlier, that the nucleosynthesis occurred over an extremely

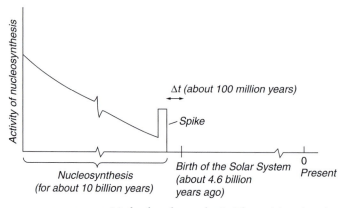

FIGURE 2.1 Mode of nucleosynthesis. The activity of nucleosynthesis or nucleosynthetic reaction decreased exponentially except for a last sudden 'spike'. Δt indicates the time between this spike and the birth of the Solar System.

long period. Iodine-129 and xenon-129 only point to a relatively recent nucleosynthetic event just before the formation of the Solar System (Figure 2.1). An even sharper constraint was obtained in the 1970s by Gerald Wasserburg and his colleagues at the California Institute of Technology (Caltech), who found evidence for the former presence of aluminum-26 (^{26}Al) in some meteorites.[2] The half-life of ^{26}Al is only about 700 thousand years, so this finding indicates that a spike-like nucleosynthesis may have occurred immediately prior to the birth of the Solar System. A possible cause for such a spike is widely considered to be the explosion of a nearby supernova. The shock of the explosion could have triggered the contraction of a molecular cloud to form the presolar nebula from which the Solar System was born. Thus, this idea explains naturally why the synthesis of extinct isotopes and the formation of the Solar System are nearly coincidental.

CHEMICAL COMPOSITION OF THE SOLAR SYSTEM

The extinct isotopes such as ^{129}I and ^{26}Al in meteorites tell us that these meteorites formed immediately after those isotopes were created. So far we are implicitly assuming that these meteorites were formed at the same time as the formation of the Solar System at large. As meteorites account for only a trivial fraction of the Solar System, one may wonder if it is safe to rely so heavily on meteorites. In fact, there are two good reasons that meteorites play an extremely important role in understanding how the Solar System formed. One is based on the chemical similarity between meteorites and the Sun, and the other is based on the physics of planetary formation. We will explain the chemical reason first.

The chemical composition of the Sun can be estimated from spectral observation of its photosphere. Since the photosphere is not affected by the ongoing nuclear fusion made possible by the enormous pressure and temperature at the center of the Sun, it is widely accepted that the photosphere represents the initial average chemical composition of the Sun. The Sun accounts for 99.87% of the mass of the Solar System, so the composition of the photosphere can be regarded as the average composition of the entire Solar System. What may be surprising is that the chemical composition of a certain kind of meteorite is very similar to the composition of the photosphere. In particular, the chemically most primitive meteorites, classified as carbonaceous chondrite type CI, have an almost one-to-one correspondence with the photosphere (Figure 2.2). Good correspondence is not seen for some elements such as hydrogen, carbon, and rare gases, because these elements are "volatile", that is, unlikely to be trapped in solid rocks. Elements that tend to stay in a solid phase are instead known as "refractory". At any rate, this remarkable chemical similarity between the Sun and carbonaceous chondrites is the foundation of cosmochemistry. For refractory elements, chondrites can represent the bulk Solar System and even bulk planets, assuming that the presolar nebula was chemically reasonably well mixed.

The spectral measurement of the photosphere's composition becomes uncertain for elements with minute concentrations, and chondrites are depleted in volatile elements. So the chemical composition of the Solar System has to be estimated from both sources of information, but this is not sufficient. As the spectral signal of three heavy rare gases

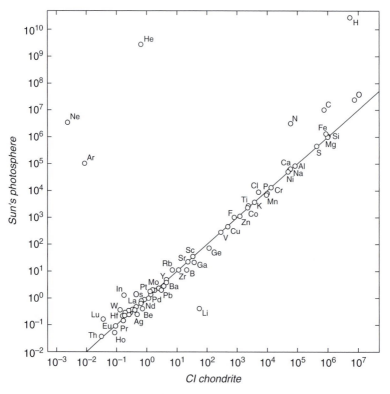

FIGURE 2.2 Elemental abundance: the Sun's photosphere versus meteorites. [9] The chemical composition of the most primitive carbonaceous chondrites is virtually identical to the Sun's photosphere, except for volatile elements. All elements are normalized to 10^6 atoms of Si. A recent spectroscopic study [35] suggests lower abundance of C, N, and O in the Sun, and thus these data will lie closer to the correlation line.

(argon, krypton, and xenon) is too weak to observe, the abundances of these elements are estimated from neighboring elements with similar atomic numbers by making use of an empirical relation between the atomic number and the elemental abundance (Figure 2.3). Solar abundances decrease nearly exponentially from light elements to heavy elements (note that the odd and even atomic numbers must be viewed separately). Suppose we attempt to estimate the abundance of argon. By joining the points corresponding to the elements whose even atomic numbers fall on both sides of argon in Figure 2.3 (sulfur, which has an

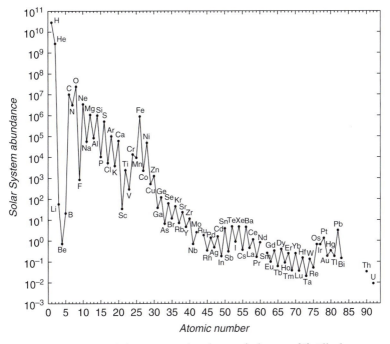

FIGURE 2.3 Solar system abundance of elements.[9] All elements are normalized to 10^6 atoms of Si.

atomic number of 16, and calcium, whose atomic number is 20, their element abundances already being known from meteorites and other means), we can obtain the value from the point on this line corresponding to the atomic number 18 of argon.

The composition of the Solar System used these days (Figure 2.3) is based on the combination of these three methods: (i) analysis of CI carbonaceous chondrites is used for refractory elements; (ii) spectral line analysis of the Sun's surface is used for volatile elements (hydrogen, carbon, nitrogen, etc.) and light rare gases (helium, neon); and (iii) the empirical relation between the atomic number and elemental abundance is used for heavy rare gases (argon, krypton, xenon). Though meteorites are just small pieces of rock fallen from the sky, they contain valuable information about the bulk Solar System. As we will explain in Chapter 4, the oldest age of meteorites can be estimated as about

4.6 billion years, based on their isotopic compositions. But how is this age related to the age of the Solar System? To answer this question, we need to know how the Solar System emerged from the presolar nebula.

FROM A GASEOUS NEBULA TO PLANETESIMALS

All of the ingredients for the Solar System are believed to have been initially spread throughout space as gas and fine dust, called the interstellar medium. The interstellar medium is extremely tenuous and hot, but eventually settles down to form clumps of cold and dense molecular cloud, each clump typically containing millions of solar masses. Such clumps are gravitationally unstable, so they tend to contract quickly and form stars. Our Solar System is thought to have been born in such an environment. If we suppose that the molecular cloud is rotating slowly in its early stages, parts of it will be unable to contract directly toward the center because of centrifugal force and will gradually flatten, eventually forming a disk called the proto-planetary disk. The temperature of the disk then falls, so the gas further condenses and the disk comes to contain innumerable small particles, whose sizes range from under a micrometer to a few milli-meters. As the density of the disk continues to increase under its own gravity, it becomes dynamically unstable and can break up into par-ticle clusters approximately a few tens to a hundred kilometers in size, which form planetesimals.

The prediction of the division of a gaseous nebula into planet-esimals is based on dynamic calculations presented at almost the same time by V. S. Safronov of the former Soviet Union, C. Hayashi of Japan, and P. Goldreich and W. R. Ward of the United States.[3] This process was called the fragmentation of the proto-planetary disk, and came to be regarded as one of the most fundamental processes in planet for-mation. However, the mathematical analysis at that time was based on a static and homogeneous nebula without considering the Sun's gravitational attraction. Although the concept of the formation of planetesimals, the building blocks of planets, was truly a breakthrough in planet formation theory, the latter assumption is obviously too simplified to be realistic. The fragmentation theory was initially criticized on the grounds that turbulent motion in the nebula would have prevented the formation of planetesimals. However, recent theoretical calculations have shown, by taking into account the Sun's gravity, that turbulent motion in the nebula could actually result in the formation of even larger planetesimals of a few hundred kilometers in size.[3] This completes the first stage of planet formation in the currently prevailing standard model.

The planetesimals thus formed then repeatedly collided with each other and combined, and finally developed into the rocky planets Earth, Mars, and so on. Dynamic calculations indicate that the process, from the presolar nebula to the planets, would have taken from several million to several tens of millions of years.[4] Meteorites are fragments of these planetesimals, so their age should not differ from the beginning of the Solar System formation by more than 100 million years. This is consistent with the view that a nearby supernova explosion resulted in the nucleosynthesis of extinct isotopes such as ^{129}I as well as the contraction of the presolar nebula at the same time.

FROM PLANETESIMALS TO PLANETS

As soon as the countless kilometer-sized or larger planetesimals were formed, they would have started to interact by gravity and grow by collisions, and it is believed that the whole system rapidly evolved into the so-called "runaway growth" phase. The concept of the runaway growth of planetesimals is the second major breakthrough in the planet formation theory. Our understanding of this concept owes much to the computer modeling of George Wetherill and his colleagues at the Carnegie Institution of Washington.[4] Bigger objects have greater gravitational attraction than smaller objects, and this advantage allows them to grow even bigger. However, the rate of growth eventually slows down, because the supply of materials runs low as the dust and smaller bodies progressively get swept into a few larger bodies. This phase of growth at a decreased rate is called the "oligarchic-growth" phase.[5] The end result of the second phase is a few dozen Moon- to

Mars-sized planetary embryos, orbiting around the Sun as "oligarchs". This second phase is estimated to take roughly 0.1 to 1 million years at the heliocentric distance of the Earth from the Sun. Gaseous proto-planetary disks around young stars are observed to dissipate within 1 to 10 million years, so the oligarchic growth would proceed before the disk gas disappeared.

The oligarchic state, with a few dozen planetary embryos orbiting around the Sun, is dynamically unstable because the planetary embryos will perturb each other gravitationally into crossing orbits leading to giant impacts. The result of this stage is the formation of full-sized planets. This is a highly stochastic process, and computer simulations show that our Solar System is just one outcome of many possibilities; different planetary configurations are equally possible with the same mass of the Solar System.[4] Computer simulations also show that this last stage takes 10 to 100 million years at the Earth's distance. The Moon is thought to be a by-product of the very last giant impact on the growing proto-Earth. The formation of the Moon is thus an integral part of the proto-planetary disk evolution. The energy involved in giant impacts is so large that both the proto-Earth and the impactor could have been entirely molten for an extended period, and the Moon is believed to have formed by the solidification and accretion of ejected materials around the Earth.[5]

CHEMICAL COMPOSITION OF THE EARTH

According to the planetary formation theory just described, the Earth formed in the first 100 million years of the Solar System history, and its composition is expected to be similar to the solar or chondrite composition at least for refractory elements. Can we test these theoretical predictions from what we observe on Earth? Let us first consider the composition of the Earth. How can we measure the average composition of the entire Earth? If you think of doing so, you may quickly realize that it is a formidable task. Samples collected by geologists from the surface of Earth's crust represent only a tiny fraction of the Earth. Also, even with current technology, it is extremely difficult to collect underground rock samples directly by drilling from depths of more than several kilometers. Mantle rocks occasionally get trapped in ascending magma and are brought to the surface as "xenoliths" (meaning foreign rocks), but even these originate from depths of less than 200 kilometers or so. Needless to say, there is virtually no possibility in the near future of obtaining direct samples from the core, which is more than 3000 kilometers down. So there is little hope of figuring out the composition by directly averaging samples from the various parts of Earth. A brute force approach clearly does not work, and we need to be a bit more creative.[6]

One important observation is that the composition of mantle rocks such as mantle xenoliths varies from sample to sample, but the variation is not random. For example, if one mantle rock happens to have more magnesium than other mantle rocks, it is likely to have less calcium than the latter. In other words, the calcium content of mantle rocks is negatively correlated with their magnesium content, and this type of systematic variation can be seen for a number of other elements as well. Why do these rocks show such chemical trends? The most plausible explanation is that the rocks have experienced different degrees of melting in the past. When a mantle rock melts, the melt phase usually has less magnesium and more calcium than the original rock. After the melt phase has escaped to somewhere else, therefore, a leftover mantle rock will be more enriched in magnesium and more depleted in calcium. If we accept this explanation, then, the most primitive mantle, that is, a special kind of mantle with no previous melting episode, should exist along the chemical trend exhibited by mantle rocks. This special kind of mantle is called the primitive mantle or the bulk silicate Earth, and it is parent to all the kinds of silicate rocks that exist in the crust and the mantle. So even though directly averaging the variable compositions of mantle samples would not give us a reliable estimate of the average mantle composition, we may derive the average composition of the crust and mantle together, by carefully studying the variability of mantle compositions.

Where exactly does the primitive mantle exist along the chemical trend? If melting is the sole process that has happened to mantle rocks, we may look at the most primitive end of the trend, for example the end with the highest calcium content, but reality is a bit more complex because the opposite of melting can also happen. As mentioned in Chapter 1, the crust is the solidified product of mantle melting, and part of the crust can return to the mantle by subduction in plate tectonics. By mixing with subducted crustal rocks, mantle rocks can be, for example, more enriched in calcium than the primitive mantle. This reverse process – called fertilization – makes our task of finding the primitive mantle more difficult than originally hoped. As a way out, we approach this problem from the angle of hypothesis testing. We can make up a putative bulk silicate Earth by removing metallic components from carbonaceous chondrites, and see whether such a putative composition exists along the chemical trend of mantle rocks. The answer turns out to be yes; that is, all known mantle samples can be *considered* to have been derived originally from chondritic materials. We can assume that the Earth was made from carbonaceous chondrites, and this assumption does not lead to any major contradiction. One of the estimates for the chemical composition of Earth, its mantle and metallic core is shown in Table 2.1.

We need to live with this rather indirect conclusion on the composition of the Earth; obviously, we cannot verify the assumption in a more straightforward way by measuring chemical compositions at every cubic kilometer within the planet. It is interesting that we are less certain about the composition of the planet we live on than the composition of the Sun, which is more than 100 kilometers away. This is because the Earth is internally differentiated and our sampling is limited to shallow depths; chemistry makes planetary business rather complex. We tend to follow the simplest assumption for convenience, but it is always important to keep in mind the difference between facts and assumptions.

AGE OF THE EARTH

We have described in Chapter 1 that the Earth is a living planet with a heat source in its interior. Mountain-building and tectonic movements

Table 2.1 *Chemical composition of Earth (weight percent)*

	Continental crust	Oceanic crust	Mantle	Core
SiO_2	60.6	49.0	45.0	
TiO_2	0.7	0.9	0.2	
Al_2O_3	15.9	16.8	3.5	
FeO	6.7	7.9	8.0	
MgO	4.7	11.6	39.5	
CaO	6.4	11.9	2.8	
Na_2O	3.1	1.9	0.2	
K_2O	1.8	0.05	0.02	
Fe				85.5
Ni				5.2
Si				6.0
S				1.9

Continental crust composition is based on the compilation of [31].
Estimating the bulk composition of oceanic crust by averaging rock samples
turns out to be tricky,[32] and what is shown here is taken from theoretical
predictions based on mantle melting experiments.[33] The "mantle"
composition in the table refers to the composition of the bulk silicate Earth
(also known as primitive mantle), which is the combination of the
continental crust and the present-day mantle, as estimated by [29]. The
composition of the core, in particular regarding light elements, has been
controversial, and the composition recommended by [34] is shown here.

have occurred continually since its birth. At present, there are no
known rocks formed at the birth of the Earth that are still preserved
on its surface. If it were possible to obtain such rocks, we would be able
to determine accurately their age – that is, the age of the Earth – by
applying radiometric dating for absolute age determination, a method
which we will describe in more detail in Chapter 4. The oldest rock we
have been able to find so far is about 4 billion years old. It is old, but

there is still a large gap, about 600 million years, between the age of the oldest rock on Earth and the age of the Solar System as determined from meteorites. This dark period of Earth's history without any preserved rock record is called the Hadean era.[7]

Our current estimate of the Earth's age is deduced indirectly from the age of meteorites. As mentioned earlier, the Earth and meteorites can be considered to have been born roughly at the same time. This conclusion is firmly supported by the evidence that both Earth and meteorites used to contain short-lived radioactive nuclides (such as ^{129}I, whose half-life is about 16 million years) which have now totally decayed to their daughter isotopes (such as ^{129}Xe, in the case of ^{129}I). These former short-lived nuclides in planetary objects clearly indicate that the birth of these objects in the Solar System must have commenced within a few tens of millions of years of the origin of the Solar System, or almost at the same time in the planetary time scale.

The age of the Earth has been pursued throughout almost the entire history of science, as one of the most fascinating problems to human beings. However, it was only in the middle of the twentieth century that we first had a meaningful quantitative answer to this problem. By elaborating the earlier seminal calculation by E. K. Gerling in Leningrad (now St Petersburg), which was the first attempt to estimate the age of the Earth based on lead isotopes, it was in 1946 that Arthur Holmes in Edinburgh and Fritz Houtermans in Bern independently published their mathematical calculations of the age of the Earth. Incidentally, Gerling was also the first scientist to discover noble gases trapped in meteorites, which opened a way to modern meteoritics (the study of meteorites).

HOLMES, HOUTERMANS, AND PATTERSON

Holmes and Houtermans both looked at lead ore mines, such as in Sudbury in Canada and Broken Hill in Australia, which are characterized not only by their immense scale of galena (lead sulfide, PbS) ore deposit, but also by their remarkable isotopic homogeneity of lead (at least to the detection level at that time). They assumed that these lead ores had been extracted from a wide-ranging volume of the Earth's interior and deposited as galena in the crust. The isotopic composition of lead in the ores was therefore assumed to be equivalent to the average isotopic composition of lead in the Earth, which was inherited from the solar nebula.

After the birth of the Earth, the lead isotopic composition continued to change through the addition of lead isotopes decayed from uranium contained in the Earth. At some later time, lead was extracted from the solid Earth to form a galena ore. Once deposited as galena, the isotopic composition of lead no longer changed because galena contains almost no uranium. We can then relate the Earth's age to the ore formation time as well as to the U/Pb ratio in the Earth using the radioactive decay equation.

A breakthrough in resolving the age of the Earth was the use of the uranium decay system. Natural uranium has two radioactive isotopes, ^{238}U and ^{235}U, which decay to ^{207}Pb and ^{206}Pb, respectively. Thanks to these two parallel isotope systems, we can resolve the age of the Earth just by measuring lead isotopic ratios, once the ore formation time has been determined by a conventional radiometric dating method. Using data from ten large galena deposits, Houtermans deduced the age of the Earth to be 4.5 billion years with an uncertainty of 300 million years.[6][8] A similar result was obtained by Holmes with a different set of galena ores.[7] Later, Clair Patterson of the United States applied the same method to meteorites.[8] Assuming that meteorites were endowed with the same initial lead isotopic composition from the solar nebula from which all planetary objects were derived, Patterson deduced the age of the meteorites to be 4.5 billion years with an uncertainty of 70 million years. The amazing agreement of this meteorite age with the age of the Earth estimated by Holmes and Houtermans, together with the common existence of extinct isotopes in both objects, has led to a consensus that the Earth was born 4.5 billion years ago.

However, we should be a little cautious here. Even the largest galena deposit may not represent the average lead isotopic composition of the Earth, and hence the agreement with the meteorite results is at best approximate. In fact, more precise analyses of numerous galena ores from all over the world have shown considerable diversity, and we should take the age of 4.5 billion years as an approximation. As we try to focus more sharply on the age with a more precise isotopic method, the image of the age of Earth is inadvertently being blurred. We have to be content with this somewhat vague definition of 'the age of the Earth'.

NOTES

1. A classic paper on stellar nucleosynthesis is [10] and a standard textbook on this topic is [11]. A good qualitative explanation can be found in [12].
2. Classic papers on this type of calculation are [13, 14]. Later developments were summarized in [15].
3. Safronov's work was originally published in Russian.[16] The work of Hayashi's group was published in English but in Japanese society journals [17, 18] and remained relatively unknown to the west for a while. The English translation of Safronov's work was acknowledged in [19].
4. George Wetherill initially worked in the field of isotope geochronology and made important contributions there. Inspired by Safronov's work, he later became interested in studying the origin of the Earth using numerical simulations and made seminal contributions here as well.[20–22]
5. A good review on the dynamics of the Moon-forming giant impact and its aftermath is [23]. It must be emphasized, however, that a giant impact is not the only viable mechanism for lunar formation.[24]
6. Geochemists have tried to estimate the composition of the Earth's mantle by a variety of methods.[25–28] A review of these attempts can be found in [29].
7. Earth's history is divided into four eons: the Hadean (4.6–4.0 billion years ago), the Archean (4.0–2.5 billion years ago), the Proterozoic (2.5–0.54 billion years ago), and the Phanerozoic (0.54 billion years ago to present). As explained in the text, the boundary between the Hadean and Archean is defined by the age of the oldest rock found on Earth. The Archean–Proterozoic boundary is defined by the relative abundance of rocks; rocks of Archean ages are much rarer than those of younger ages. The Proterozoic–Phanerozoic boundary is defined by the emergence of abundant animal life; the name Phanerozoic derives from Greek words meaning "evident life".
8. One of the authors of this book (Ozima) had an opportunity to meet Fritz Houtermans, and this encounter was quite memorable. One afternoon in the fall of 1957, Ozima's thesis-supervisor Don Russell brought a guest to an

underground laboratory at the Physics Department of the University of Toronto, where Ozima was constructing a K–Ar radiometric dating line. The guest's very first greeting was, "May I have your cigarette?" Ozima was stunned, and even more surprised to see that the distinguished-looking visitor had almost no front teeth. The guest was Professor Fritz Houtermans, who had come to attend the International Union of Geodesy and Geophysics meeting in Toronto. Shortly after, Ozima learned that when Houtermans was visiting the Soviet Union to work on physics, he was arrested under the allegation of being a German spy and imprisoned for almost three years. Owing to cruel physical tortures during interrogation, he lost most of his teeth. However, thanks to his wife and many eminent friends including Professors von Laue and Landau, he was finally released and sent back to Nazi Germany. Even under threats to his life during interrogation, he never compromised his friends. After the war, he moved to Bern in Switzerland to set up a small laboratory, which eventually became a center of astrophysical research. The incredible record of his fight for survival in the prison is a hope and triumph for man (see the July 1992 issue of *Physics Today* for a brief but moving account of his unique experiences [30]).

3 Formation of the layered structure of the Earth

The most distinctive feature of the Earth's interior is its chemically layered structure. If we were to cut the planet in half, we would find a structure similar to that of an onion (Figure 3.1). The first 3500 km or so from the center of the Earth is called the core, and its main components are thought to be iron and nickel, with an additional 10 percent consisting of a mixture of lighter elements. There are various theories for what these light elements could be. Some researchers believe sulfur to be present, while others have suggested silicon or oxidized iron, but no conclusion has been reached yet. It is known that the core is further divided into two sub-layers. The inner core is a solid with a radius of about 1200 km, while the outer core is in a liquid state.

The layer that extends from the outer core toward the Earth's surface is called the mantle, and it is generally considered to have a composition similar to that of peridotite. The outside of the mantle, i.e. the surface layer of the Earth, is called the crust. On the continents the crust is 40 to 50 km thick, but under oceans it is only several kilometers thick.

Most of our knowledge about this basic layered structure is gained from how seismic waves propagate through the Earth and from laboratory experiments on the state of matter under high pressures and temperatures; pressure at the center of the Earth reaches 3 million atmospheres, and the Earth's internal temperature is a few thousand degrees centigrade.

At what stage in the history of the Earth was this layered structure formed? Was it formed at the same time as the Earth was born? The formation of the layered structure is undoubtedly one of the most

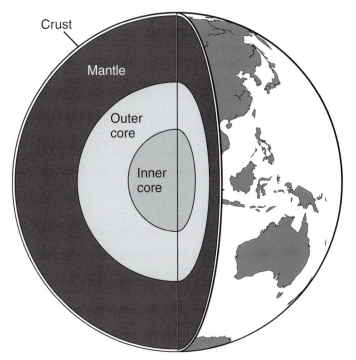

FIGURE 3.1 Layered structure of Earth's interior. The crust and mantle are made of silicate rocks while the core is made of iron–nickel alloy.

fundamental issues in Earth sciences, and as we will see in the following, it is a very challenging problem.

HOMOGENEOUS AND HETEROGENEOUS ACCRETION HYPOTHESES

It was once thought that the Earth was born first and then the core was formed. In this hypothesis, tiny planetesimals accumulated and formed a homogeneous primordial Earth. The temperature of the planet then gradually rose owing to the energy from gravitational contraction and the nuclear disintegration of radioactive elements, and iron and nickel were separated from the primitive silicate materials of which the primordial Earth was composed. Iron and nickel then melted and, because of their high density, fell down into the center of the Earth and formed the core. This is known as the homogeneous accretion hypothesis.[1]

However, this hypothesis leaves many points unexplained. For example, even though it is reasonable to expect the melting of iron and nickel in the primordial Earth because they have lower melting points than silicates, it is unclear if they could really sink to the Earth's center, traveling over a distance of more than 1000 km. Another problem is oxidation. Iron and nickel can melt out as metal in a very reduced environment, so for the homogeneous accretion hypothesis to work, the primordial Earth must have been very reduced. However, the present-day mantle, which is the leftover component after the core was segregated, is much more oxidized. One plausible way to oxidize the mantle is to react with water, but the required amount of water is as much as 60 present-day oceans. Such oxidation also releases a huge amount of hydrogen, which must have been lost into space quickly, to be consistent with the composition of the atmosphere over Earth's history. Finally, mantle rocks we sample today contain a few thousand ppm of nickel. Though this may seem a trivial amount, virtually no nickel should exist in the mantle according to the homogeneous accretion hypothesis, because almost all nickel should go into the core during the melting of the primordial Earth; nickel has such strong chemical affinity to iron.

For these reasons mentioned above, some researchers have suggested that the core may have formed at the same time as the Earth. According to this idea, the Earth has possessed the basic layered structure of core and mantle since its birth. This is called the heterogeneous accretion hypothesis, which proceeds as follows.[2] When a cold molecular cloud contracts under its own gravity, its temperature rises because of the release of gravitational energy, and the temperature of the initial presolar nebula is estimated to have been as high as several thousand degrees kelvin. The nebula then cooled gradually owing to the radiative loss of heat. The presolar nebula consisted mostly of hydrogen and helium; its chemical composition was the same as the solar composition discussed in Chapter 2. When the temperature of the presolar nebula was high, various atoms could move around at random, but as the temperature dropped, atoms bonded chemically to form more stable molecules, and then condensed and formed fine particles. Condensation would have occurred generally in order, beginning with those molecules that have the highest melting points.

The order in which different molecules may have condensed can be predicted in considerable detail from thermodynamics. This prediction is the so-called 'condensation theory'.[3] Thermodynamic calculations have shown that if presolar nebulae cooled slowly enough to maintain chemical equilibrium, the first molecules to condense would have been the oxides of aluminum and calcium (e.g. Al_2O_3 and $CaAl_{12}O_{19}$), followed by iron and nickel. These molecules condense at temperatures between 1700 and 1400 K. When the presolar nebula cooled below about 1400 K, magnesium silicates condensed. As the temperature continued to fall, sulfides, lead chlorides, and iron oxides condensed. When the nebula cooled to room temperature, nearly all of the molecules had condensed, including water.

Condensed particles collided with each other and formed larger and larger particles by repeated collision. Let us suppose that the formation of planets occurred at the same time as the formation of these fine particles; that is, as soon as fine particles condensed from the nebular gas, they combined and began to form planets. The fine particles formed in the initial stage of condensation would accrete first and form the central part of the Earth. As mentioned above, the fine particles that would condense at the beginning are magnesium and calcium oxides, followed by iron and nickel. So in this process of accretion, the aluminum and calcium oxides, followed by iron and nickel, would form the Earth's center, and silicates would accrete around them, with the outermost layer being volatile elements, including water. So the layered structure of Earth could have been formed through heterogeneous accretion.

The above concept may explain the formation of a certain kind of layered structure in a planet, but it is based on the assumption that condensation and accretion occurred simultaneously. From dynamic calculations on planetary formation, however, it is difficult to expect such rapid and well-ordered formation. And above all, the aluminum and calcium oxides, which have the highest melting points, would have to be in the center of the Earth, contrary to our observations.

CHEMICAL DIFFERENTIATION AND PLANETARY ACCRETION

The homogeneous and heterogeneous accretion hypotheses both illustrate that resolving the origin of Earth's layered structure, which might seem a very simple problem, is not straightforward at all. These hypotheses appear to be too simple to be a narrative for what has actually shaped the Earth. Both ideas were proposed many years ago, and since then a variety of models have been suggested.[4] One important notion that has emerged in the discussion is that the segregation of the core from primordial materials is unlikely to be a single event. As soon as planetesimals reached a certain size, they seem to have become hot enough to melt and segregate metallic components. We have learned this from studying meteorites.

In the preceding chapters, we focused on a special kind of meteorites called carbonaceous chondrites, but there are many other kinds of meteorites, which can be divided broadly into three groups: stony, iron, and stony-iron, the last being somewhere between the first two. Carbonaceous chondrites belong to the group of stony meteorites and are considered to be most primitive, representing primordial materials condensed from the presolar nebula. Other types of stony meteorites exhibit the trace of chemical differentiation; that is, parental bodies for those meteorites must have experienced melting. Moreover, iron and stony-iron meteorites are best explained as the product of melting. They are fragments of planetesimals that have melted and formed a metallic core.

When planetesimals were growing by collision, therefore, each of them probably already had a metallic core inside, and a core grew at the same time as a planet grew. As explained in Chapter 2, the final stage of planetary formation is through giant impacts, and it is widely believed that the very last giant impact resulted in the creation of the Earth–Moon system. The energy involved in a giant impact is considerable, and theoretical calculations suggest that, right after the Moon-forming giant impact, a substantial fraction of the Earth was in a molten state, which is called a "magma ocean". In such a globally molten Earth, metallic components would have been easily aggregated, and it is now considered that the core's formation was finalized at the birth of the Earth–Moon system.

IRON CORE AND SILICATE MANTLE SEPARATION

Planetary formation from tiny planetesimals through giant impacts is a highly stochastic process, and it is impossible to delineate how exactly the Earth's core grew during planetary accretion. Using isotopes, however, we can still resolve the time scale involved in the core formation. A key radioactive isotope here is hafnium-182 (^{182}Hf), which decays to tungsten-182 (^{182}W) with a half-life of 9 million years. As is common practice in dealing with isotopes, we use an isotopic ratio rather than absolute concentrations, which frees us from bothering with scales including mass and volume when discussing the geochemical behavior of isotopes. Instead of the absolute concentration of ^{182}Hf and ^{182}W, we use ^{182}Hf/^{180}Hf and ^{182}W/^{184}W, in which ^{180}Hf and ^{184}W are stable isotopes.

We can trace core–mantle segregation with the hafnium–tungsten system because hafnium and tungsten have very different chemical affinity to iron. When segregation of iron from silicate takes place, tungsten, which has strong affinity to iron, goes with iron into the iron core, but hafnium remains essentially in the silicate phase. Therefore, after the core formation, the ratio of hafnium to tungsten Hf/W is high in the silicate mantle, whereas the iron core is practically deprived of hafnium. This elemental fractionation between hafnium and tungsten affects the subsequent isotopic evolution of tungsten, because while radiogenic ^{182}W decayed from ^{182}Hf accumulates in the silicate mantle, the tungsten isotopic composition remains practically unchanged in the iron core.

Because of its very short half-life, ^{182}Hf no longer exists in the Solar System, but its trace is left in the present-day tungsten isotopic ratio ^{182}W/^{184}W, from which we can extract the time scale of iron–silicate segregation. The Hf–W chronometer may be likened to a sand-glass that ceased to work a long time ago. In comparison, common radiometric chronometers such as the potassium–argon or uranium–lead system are analogous to a clock with two hands, which is still ticking.

HAFNIUM–TUNGSTEN CHRONOMETER: A GEOLOGICAL SANDGLASS

In Figure 3.2, we show schematically the evolution of the tungsten isotopic ratio $\left(^{182}W/^{184}W\right)$ in a diagram of $\left(^{182}W/^{184}W\right)$ versus $\left(^{180}Hf/^{184}W\right)$. Here, for the sake of illustration, we consider a collection of planetary objects that initially shared the same isotopic ratio of ^{182}W/^{184}W but have different elemental ratios of Hf/W (i.e. different ^{182}Hf/^{184}W). In this diagram, these objects should lie on a straight line because an object with a higher initial Hf/W will have a higher

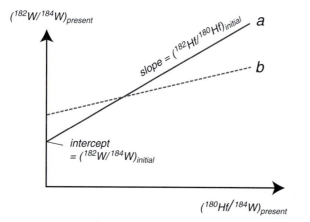

FIGURE 3.2 Schematic illustration for the evolution of the Hf–W isotopic system. In this space of the present values of ^{182}W/^{184}W and ^{180}Hf/^{184}W, objects that share the same initial ^{182}W/^{184}W lie on a straight line, the slope of which is identical to the initial ^{182}Hf/^{180}Hf. Data from primitive chondrites form a steeper line (a) than those from the Earth–Moon system (b).

present-day $^{182}W/^{184}W$ (because ^{182}Hf decayed to ^{182}W) as well as a higher $^{180}Hf/^{184}W$ (because this ratio of the stable isotopes will remain the same). The slope of this straight line is proportional to $^{182}Hf/^{180}Hf$. If there were still a substantial amount of ^{182}Hf when those objects were formed, the slope would be steep, but if all of the ^{182}Hf had already decayed, the straight line would be flat because there is no reason for $^{182}W/^{184}W$ to change with time.

Once the presolar nebula was isolated from the interstellar medium in the galaxy, ^{182}Hf was no longer produced in the Solar System, and $^{182}Hf/^{180}Hf$ only decayed monotonically following the law of radioactive decay. Therefore, the $^{182}Hf/^{180}Hf$ ratio yields an absolute measure of time since the sequestration of the presolar nebula. Primitive meteorites form a steep linear array in the tungsten isotope diagram because they formed shortly after the formation of the presolar nebula and had a high $^{182}Hf/^{180}Hf$ ratio. Other planetary objects that formed at a later time were endowed with a lower $^{182}Hf/^{180}Hf$ ratio, so they should form a less steep linear array. Also, the intercept of the array with the vertical axis (the initial ratio of $^{182}W/^{184}W$) must be greater in accordance with a rise in $^{182}W/^{184}W$ in the presolar nebula due to the decay of ^{182}Hf.

In Figure 3.2, laboratory data obtained on primitive meteorites correspond to a steeper line, whereas those for the Earth–Moon system lie on a less steep line, indicating that the Earth–Moon system was created after the formation of primitive meteorites. The positive slope of the Earth–Moon system is very important. The formation of the Earth–Moon system by a giant impact resulted in two isolated silicate parts (Earth's mantle and the lunar mantle) with different Hf/W ratios, but if the formation took place after all ^{182}Hf had decayed into ^{182}W, the $^{182}W/^{184}W$ ratio of those mantles would be the same, which is equivalent to having a zero slope in the tungsten isotope diagram. Using the law of radioactive decay, we can estimate that the formation of the Earth–Moon system took place about 30 million years after the formation of the primitive meteorites. The tungsten isotopes thus yielded the first convincing information on the timing of mantle–core separation in the Earth–Moon system, which was reported in 2002 independently by three groups of scientists in the United States, Germany, and Australia.[1–3]

FORMATION OF THE INNER CORE

From seismology, we know that the iron core now consists of two parts, a liquid outer core and a solid inner core. One may naturally wonder when and how the inner solid core formed. The issue is not only about the basic structure of the Earth; it also has profound implications for the long-term evolution of the Earth, in particular for the origin of the geomagnetic field.

The geomagnetic field is generated by vigorous fluid motion in the outer core, which can be likened to a huge electric generator in a power station. We will discuss this more in Chapter 9. Some researchers suggest that the existence of a substantial inner core is needed to stabilize a stationary geomagnetic field as we now observe. This idea is still hotly debated,[4] but if it is true, it would mean that full development of the geomagnetic field must postdate the formation of an inner solid core.

Soon after the Moon-forming giant impact, the temperature of the Earth–Moon system was so high that the whole system was very likely in a molten state, where silicate and iron separation proceeded to form a molten iron core. As the temperature fell further, the center of the core where the pressure is the highest first dropped below the melting point of iron. This is expected from the extrapolation of laboratory experiments on the melting temperature of iron at high pressures. With further cooling of the Earth's interior, the inner core is expected to grow, but the long-term thermal evolution of the inner core must be understood in the framework of the thermal evolution of the whole Earth, in which the delicate balance between heat production from radioactive decay and heat loss from the surface becomes important. Researchers are still debating when in the Earth's history the solid inner core started to form and how quickly it grew to its present size.[5]

THE EARTH'S CORE AND THE GEOMAGNETIC FIELD

The Earth has a magnetic field of about 30 000 nanoteslas at the surface. For comparison, the strength of a typical refrigerator magnet is a few milliteslas, which is a hundred times stronger than the geomagnetic field. The geomagnetic field thus appears to be very weak, but it is also of planetary scale. The spatial pattern of the field is very similar to the so-called dipole field, as if it were produced by a bar magnet placed at the center of the Earth and parallel to its rotation axis; such a hypothetical bar magnet has to have an enormous strength because the dipole field strength decreases with the cube of the distance from the source. The question of how the Earth's magnetic field is generated is one of the most fascinating problems in geophysics, and many theories have been put forward, all of which agree that it is generated by flows in the liquid part of the core. The movement of the fluid body of iron and nickel functions as a generator and continuously sends electric currents through the liquid to form a magnetic field. This idea is generally known as the geodynamo theory, which was first proposed by E. C. Bullard of Cambridge University in 1949.[6]

The fact that the generation of the Earth's magnetic field is closely linked to the movements of the liquid part of the core can also be understood from the observations of the variability of the magnetic field. At first sight, the Earth's magnetic field seems to be more or less stationary, but on a closer look it is actually fluctuating constantly. Even in the past 100 years, during which systematic observation of Earth's magnetic field has been carried out, its strength has decreased almost linearly by 6 percent. If it continues to decrease at this pace, it will reach zero in less than 2000 years. It is difficult to understand such short-term variations if one supposes that the field is somehow created by rocks deep within the Earth, because it does not seem possible that solid rocks could move around on such a time scale.

Since the publication of Bullard's now classic paper, theories on the origin of the Earth's magnetic field have pursued the nature of the fluid motions within the core that could give rise to the magnetic field. It used to be assumed that these fluid motions were large-scale convecting movements in the liquid outer core. Later research showed, however, that it would be extremely difficult to maintain the Earth's magnetic field with only large-scale convection. It is now believed that the fluid movements maintaining Earth's magnetic field are probably small-scale turbulent flows.[7] At any rate, fluid motion in the core would not take place if the core were not cooled by the mantle; in other words, the core is thermally insulated by the mantle, and the cooling of the mantle determines the long-term behavior of the geomagnetic field. Unlike Earth, Venus and Mars lack a planetary-scale magnetic field, and this difference is usually attributed to the absence of plate tectonics on those planets. Plate tectonics, which is so far observed only on Earth, can cool a planet very efficiently, so it is probably essential for the generation of the geomagnetic field.

MANTLE, CRUST, AND ATMOSPHERE

The separation between core and mantle is only one of the major chemical differentiations that are responsible for the Earth's layered structure. We have the crust–mantle boundary and also the atmosphere above the solid Earth (Figure 3.1). As discussed above, the formation of the core–mantle system was probably finalized by the Moon-forming giant impact. In resolving the time scale of this process, we used the Hf–W chronometer. We emphasize that this Hf–W approach took full advantage of two characteristic features of this system. One is the very short half-life of ^{182}Hf as a precise time marker in the early Solar System, and another is the marked difference in elemental affinity between the two separated layers considered, i.e. silicate and iron.

The separation of the crust and the mantle is a continuous process, which is still ongoing today. We need a different radioactive isotope system with a much longer half-life as a time marker; we need a ticking clock instead of a sandglass. Also, in order to constrain the timing of the separation process, we need an appropriate combination of parent and daughter isotopes, which respond to a separation event through elemental fractionation. In Chapter 6, we will discuss the application of the samarium–neodymium (Sm–Nd) isotope system to the crust–mantle separation. ^{147}Sm decays to ^{143}Nd with a half-life of about 100 billion years, and Nd tends to be more concentrated in the crust than Sm. These two characteristics allow us to trace the separation event over the Earth's history.

The same approach can also be applied to the separation of the atmosphere from the solid Earth. It is natural to choose a radioactive isotope system in which a decay product (i.e. a daughter isotope) is a gas. The potassium–argon and iodine–xenon systems can thus be used to track the formation of the atmosphere. We will come back to this problem in Chapter 7. Before embarking on these exciting applications, however, it is important to grasp the fundamentals of various chronometers used in Earth sciences, which will be reviewed in the next chapter.

NOTES

1. Ted Ringwood was a major proponent of the homogeneous accretion hypothesis.[8] Other notable studies include [9, 10].
2. An idea on heterogeneous accretion was proposed in 1944 by Eucken.[11] More modern variations include [12–14].
3. Harold Urey also pioneered this thermodynamic approach.[9] The first detailed calculation was done by [15]. See [16] for a comprehensive review on this subject.
4. For recent reviews, see [17–19].

4 Time scale of the Earth's evolution

MEASURING THOUSANDS OF MILLIONS OF YEARS

How do we determine the time scale that forms the backbone of the Earth's evolution? This chapter will focus on the method of measuring incredibly long "geological ages", far exceeding the bounds of human experience.

Our sense of time is usually connected to some kind of change in geometrical or physical quantities. Taking a watch as an example, the angle of the hand in its revolution corresponds to the time. Similarly, when measuring geological ages of thousands of millions of years, it is necessary to find some appropriate quantity for the transition in time, such as the length, angle, or weight of an object.

Though the principle in measuring geological time is the same as in measuring time in everyday life, several important problems arise from the extraordinarily long time involved. A particularly important aspect is that there must be a guarantee that a "geological clock" has moved at the same pace for thousands of millions of years. At best, a human lifespan is no more than 100 years or so, and the history of the entire human race covers less than a million years. Even if we are successful in finding a "geological clock" and can ascertain that it has been ticking away at an extremely regular pace since we have been on Earth, how can we be sure that it has moved at the same pace over a period hundreds of thousands of times longer than our existence?

Another essential condition for a geological clock is that the measure marking time must not be influenced at all by the environment. Some volcanic rocks have been exposed to weathering for a long time since they erupted on the Earth's surface, while others, such as granite, solidified deep within the Earth and have existed under high pressures and temperatures. Obviously the clock is of little use if it is influenced by such external factors. Radioactive decay, or a change in the number of a particular isotope due to nuclear disintegration, is used as a yardstick for geological time because this approach overcomes these challenges.

GEOLOGICAL CLOCK USING NUCLEAR DISINTEGRATION
We have already said that uranium and potassium (potassium-40 to be exact) undergo radioactive decay into lead and argon, respectively. The rate of decay is specified by their particular half-lives. Uranium-235, uranium-238, and potassium-40 have half-lives ranging from 700 million to 5 billion years. In nuclear disintegration, the number of nuclei of a given radioactive isotope (N) will decrease with time following a very simple rule: the number of nuclei that disintegrate over a certain period is proportional to the total number of nuclei in the system. This proportion constant is called the decay constant and is usually written as λ. Expressed mathematically, the law of radioactive decay is $dN/dt = -\lambda N$. This formula is exactly the same as that used to find the probability of a certain number of electric bulbs burning out in a certain period of time. It is essentially a statistical law, and the stability of the internal structure of a given radioactive isotope is succinctly captured by its decay constant.

The change in the number of nuclei with time is shown in Figure 4.1a. The parent isotopes (^{238}U, ^{235}U, ^{40}K, etc.) decrease exponentially with time. Corresponding to this decrease is an increase with time in the number of the daughter isotopes formed from the parent isotopes. Figure 4.1b shows how the ratio between the daughter and parent isotopes changes with time. This ratio obviously increases as time passes. The shape of the curves in Figures 4.1a and b is determined by the decay constant, which specifies the rate of disintegration. An isotope with a larger decay constant disintegrates more quickly, corresponding to a shorter half-life.

Let us take ^{40}K as an example. Volcanic rocks such as basalt usually contain a considerable amount of potassium (usually 0.1–1 percent). It is

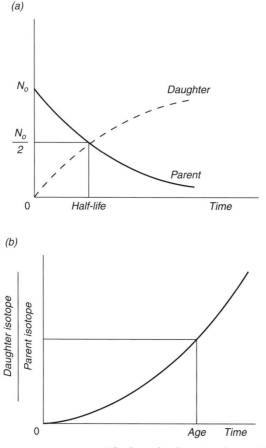

FIGURE 4.1 The law of radioactive decay. (a) Change with time of the parent (solid) and daughter (dashed) isotopes. The vertical axis shows the number of nuclei. N_0 is the initial abundance of the parent isotope. (b) Change with time of the ratio of daughter to parent isotope.

quite conceivable that, when such rocks were erupted on the surface originally as magma, nearly all of the gas contained in the magma was released. Volcanic rocks that have erupted only recently, therefore, do not contain any gas including ^{40}Ar, even if they contain ^{40}K. The ^{40}K in the rocks then gradually undergoes nuclear disintegration into ^{40}Ar. Even though argon is a gas, once volcanic rock has solidified, the argon in the minerals is trapped firmly within the crystal lattice and cannot escape. So as time passes, the ratio of the daughter isotope (^{40}Ar) to the parent isotope (^{40}K) gradually increases as in Figure 4.1b, and it is possible to find the age of the rock (value on the horizontal axis) from the ratio of ^{40}Ar/^{40}K (value on the vertical axis).

The ^{40}K and ^{40}Ar contained in the rock are analyzed in order to measure the age of the rock. As mentioned above, the amount of potassium contained in volcanic rocks is about 0.1–1 percent, and ^{40}K accounts for about 0.01 percent of this, in other words less than a millionth of the weight of the whole rock. Even after 10 million years, the amount of argon formed by the decay of this small amount of potassium would be at the most about five-billionths of a gram per gram of volcanic rock. Quantitative analysis of such minute amounts is far beyond the power of normal chemical analyses. Such analysis first became possible with the introduction of the "isotope dilution method" using a mass spectrometer, which was finally put into practical use in the late 1950s. It would not be an exaggeration to say that the development of the mass spectrometer almost parallels the progress of geological age determination, or isotope geochronology. Tracing the development of the mass spectrometer is in itself extremely interesting, but we will not go into any further detail here.[1]

CORROBORATION OF THE GEOLOGICAL CLOCK

The above is a brief account of the principle of the geological clock based on nuclear disintegration. We have seen that in the case of potassium, the age at which a volcanic rock erupted can be related unambiguously to the ratio between the parent isotope, potassium-40, and its daughter isotope, argon-40. This is known as the potassium–argon dating method, which was perfected in the 1960s and is still widely used.

We can verify in the laboratory that the disintegration of a certain radioactive isotope follows the law of nuclear disintegration, but how can we be sure that this law holds for thousands of millions of years? The backing of Earth science observations is useful in answering this question. In addition to the potassium–argon dating method, many other methods based on nuclear disintegration, such as the uranium–lead method and the rubidium–strontium method, have been suggested and are in practical use as geological clocks.[1] A large number of rocks have been dated using these methods. If we take one rock sample and date it using different methods, all of these methods will give the same age if the sample is not too weathered or altered to obtain reliable age information. This is the case whether the rock is old or young. For example, the ages of meteorites measured by the uranium–lead method, the rubidium–strontium method, and other methods all cluster around 4.56 billion years. If the law of nuclear disintegration were not valid, it would be impossible to explain the conformity of such values.

It is also known that nuclear disintegration is almost completely unaffected by the external environment. The law of nuclear disintegration, which was originally discovered at room temperature and one atmospheric pressure, still applies even under extreme conditions such as those found several hundred kilometers underground, where the pressure can exceed a hundred thousand atmospheres and temperatures reach a few thousand degrees centigrade. In other words, the pace at which the "clock" ticks away is constant.

Thus the geological clock based on nuclear disintegration is an ideal clock, which can measure incredibly long ages and is not influenced by the external environment. This law was put forward jointly in 1902 by the New Zealand-born physicist E. Rutherford and the young chemist F. Soddy from Oxford University when they were both doing research at McGill University in Montreal.[2] As said before, the law is of statistical nature, and at the time of its discovery, the decay constant was merely an empirical parameter as the stability of nuclei was not well understood. The particulars of the discovery of this law furnish a lesson in how to carry out scientific research. It is possible and indeed useful to make laws about the state of an object even if one does not have a complete understanding of it.

POTASSIUM–ARGON METHOD

Since potassium is widely distributed in nature, the potassium–argon dating method can be applied to many types of rocks. Application of the uranium–lead method is more limited, since only rocks such as granite contain enough uranium. Moreover, the half-life of potassium-40 is about a billion years, which is quite short compared with that of rubidium-87 (about 49 billion years). With rubidium-87, which is used in the rubidium–strontium method, even after 10 million years the proportion that has undergone nuclear disintegration is no more than a ten-thousandth of the original rubidium. Since this amount is extremely small, it is difficult to measure. Consequently clocks using nuclei with long half-lives are not suitable for measuring the age of young rocks. It is like measuring seconds with the hour hand of a watch. Potassium-40 is more suited to measuring the age of young rocks because of its relatively short half-life.

At present, with the exception of some special cases where the amount of rubidium is unusually large, a few million years is the minimum age that the rubidium–strontium method can measure. The potassium–argon method can measure ages of several hundred thousand years with sufficient precision. If minerals such as sanidine, which contain a particularly large amount of potassium (several percent), are used, it is possible to measure ages of several tens of thousands of years, but measuring younger ages than these is beyond the power of this method. In this case, carbon-14, which has a short half-life of 5730 years, is used. As a very successful example of the potassium–argon dating method, let us next explain how the reversals of the Earth's magnetic field were confirmed.

REVERSALS OF THE EARTH'S MAGNETIC FIELD

The Earth's magnetic field is another main actor in the planet's evolution. In Chapter 3 we discussed how this magnetic field is related to the dynamics of the core. The existence of the magnetic field can be traced back to at least 3 billion years ago.

The Earth's magnetic field at present has its magnetic south pole near the geographic north pole and the magnetic north pole near the geographic south pole; thus, the north pole of a magnetic compass points toward the geographic north pole. But the Earth's magnetic field does not always have its south pole around the geographic north pole. It is now known that the polarity changes with time.

The possibility that the polarity of the geomagnetic field might vary with time was first pointed out by B. Brunhes of France at the beginning of the twentieth century.[2] In the course of systematically measuring the remanent magnetism of volcanic rocks in Europe, Brunhes realized that several volcanic rocks were magnetized in the opposite direction to the present magnetic field. Since then, remanent magnetism facing in the opposite direction to the present magnetic field has been found in various places around the world. For example, M. Matuyama at Kyoto University discovered that some volcanic rocks with varying ages from southwest Japan and the Korean Peninsula were magnetized in the opposite direction to the present, suggesting the possibility of periodic reversals of the Earth's magnetic field.[3] Nevertheless, it was too radical an idea to conclude from these data alone, and so a further reliable proof was essential. This is where the potassium–argon method of dating volcanic rocks proved useful.

At the beginning of the 1960s, a research group called the 'Branch of Theoretical Geophysics' had just commenced research in a corner of the US Geological Survey.[3] The group's leaders were R. Doell and A. Cox, and they were joined shortly after by B. Dalrymple. All three had graduated from the Department of Geology of the University of California, Berkeley. The aim of this group was to measure the direction of the magnetism of volcanic rocks from various parts of the world, and at the same time to find their ages by the potassium–argon method. In order to carry out this program, they had to solve several formidable problems. The most difficult of these was how to determine precisely the age of young volcanic rocks several millions to several hundreds of thousands of years old. This problem was over-come by using the most accurate mass spectrometer in the world at that time, developed by J. Reynolds. As data were accumulated, the state of the Earth's past magnetic field gradually became clear. Volcanic rocks that had erupted by about 690 000 years ago were almost always magnetized in the same direction as the present magnetic field. But those erupted between 890 000 and 690 000 years ago were magnetized in the opposite direction. If the ages of volcanic rocks were the same, the polarity of their remanent magnetism was also the same, no matter from what part of Earth they were collected. The fact that all rocks of a certain age have the same polarity cannot be explained by sheer coincidence. The only plausible explanation for this phenomenon is that the magnetic field itself has reversed. So the reversals of the Earth's magnetic field were given a final proof with the help of dating by the potassium–argon method.

Since the first discovery of volcanic rocks with reversed polarity more than a century ago, there are now numerous measurements of remanent magnetism of volcanic rocks, ocean drilling cores, and sedi-mentary rocks ranging from the modern to the Precambrian period from all over the world. From the compilation of these data, we have a magnetic polarity time scale indexed with precise absolute ages measured by radiometric dating (Figure 4.2). With the use of this scale, researchers can identify the age of the ocean floor from measur-ing the magnetic polarity pattern of the ocean floor.

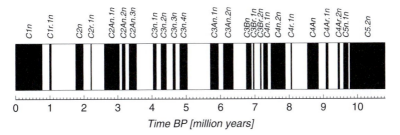

FIGURE 4.2 Example of the magnetic polarity time scale.[5] Normal and reversal polarities are denoted by black and white, respectively. Labels are polarity chrons. BP, before present.

AGE OF SEDIMENTARY ROCKS

The radiometric geological clock is the sole device to measure time for the Precambrian period since few fossils from this period exist on Earth,[4] but this does not mean that the traditional paleontological method is no longer necessary. For example, the radiometric geological clock is of practically no use with sedimentary rocks. Sedimentary rocks were formed by settling of fine particles, derived from the breakdown of various existing rocks. So if the geological clock is applied to these rocks, the age obtained will be the average value of the ages when the crystals forming the rocks were formed, and will have almost no relation to the time when these rocks formed by sedimentation. The most effective means to find the age of sedimentary rocks is by examining the fossils they contain.

Nevertheless, even if some conclusion may be drawn for the relative ages of rocks through paleontological methods, no conclusion can be made about their absolute ages. So the only way to mark some numerical values on the paleontological time scale is through the radiometric geological clock.

In estimating the age of sedimentary rocks by the radiometric geological clock alone, we must depend on the indirect method of first finding the age of volcanic rocks whose relationship with sedimentary rocks is known from geological considerations, and then inferring the age of the sedimentary rocks from this age. This fact has often been overlooked by paleontologists and geochronologists alike. We must be well aware of this point when asking, for example, how long ago the Cretaceous and Carboniferous periods were[4].

SHORT-LIVED ISOTOPE: A GEOLOGICAL SANDGLASS

So far we have discussed the use of radioactive decay as a clock (known as radiometric dating), in which the ratio of a radioactive isotope and its decayed isotope gives a measure of time elapsed since both isotopes were confined in a closed system such as a rock or mineral. In order for the radiometric dating to be useful as a geological tool, we must choose a suitable radioactive isotope that has a half-life comparable with the geological time scale of interest. In discussing the early history of the Earth, we must not only deal with the absolute ages of events close to the age of the Earth, but also resolve the sequence of such events separated by much shorter time intervals.

However, if we use isotopes with half-lives comparable with the age of the Earth, temporal resolution becomes necessarily large, because uncertainty in the measured age is proportional to its absolute age. Accordingly, in order to attain a much shorter resolution, i.e. a higher analytical precision, we must use an isotope with a much shorter half-life. The problem is that because of the shorter life, the isotope would have decayed long ago and would no longer exist in nature. So the situation is analogous to a sandglass with all the sand having run out. Can we deduce any age information from a dead sandglass like this? An ingenious way to circumvent this dilemma was shown by Typhoon Lee, Dimitri Papanastassiou and G. J. Wasserburg of the United States, who used the short-lived isotope ^{26}Al (half-life of 0.7 million years) to resolve short time intervals of a few tens of thousands of years among various types of meteorites formed near the beginning of the Solar System's history.[4]

Aluminum in nature is now a mono-isotopic element consisting of nothing but the stable isotope ^{27}Al. However, the interstellar cloud from which our solar nebula was isolated about 4.6 billion years ago is known to have contained the radioactive isotope ^{26}Al produced by nucleosynthesis. Since aluminum (including ^{26}Al) cannot be produced within the Solar System, ^{26}Al inherited from the interstellar cloud underwent one-way radioactive decay and no longer exists today. Instead, we have its decayed product ^{26}Mg. In addition to this radiogenic ^{26}Mg, current magnesium in nature has two other stable isotopes, ^{24}Mg and ^{25}Mg.

Suppose that we made a sandglass by filling the upper bulb with aluminum sands collected soon after the isolation of the solar nebula. These sands still contained ^{26}Al, which subsequently decayed to ^{26}Mg. If we observe this aluminum-sandglass in the present day, all sands are in the lower bulb, in which some sands contain ^{26}Mg decayed from ^{26}Al. If we were to repeat this experiment, but collect the aluminum sands long after the isolation of the solar nebula – for example hundreds of millions of years later – the sands in the upper bulb would contain no ^{26}Al as it would all have decayed into ^{26}Mg. In this case, no ^{26}Mg would be present in the lower half of the sandglass. Therefore, depending on when a sandglass was made and started, the isotopic composition that we now observe must be different. From this difference we can infer when the sandglass was made and started (here we assume that a sandglass is turned over at the same time it is made).

CAI: THE OLDEST MINERAL IN THE SOLAR SYSTEM

As soon as the Solar System was isolated from the interstellar cloud, ^{26}Al underwent one-way radioactive decay. Accordingly, the isotopic ^{26}Al/^{27}Al ratio has exponentially decreased and is now essentially zero. Let us again take the analogy of a sandglass. If an aluminum sandglass was made within a few tens of millions of years after the isolation of the presolar nebula, the sands must now contain some ^{26}Mg, which is the decay product of ^{26}Al. Therefore, the earlier the sandglass was made and started, the higher the ^{26}Mg/^{27}Al ratio in the sandglass will be. Since all of the ^{26}Mg in the glass was once ^{26}Al, the ^{26}Mg/^{27}Al ratio in the glass is equivalent to the ^{26}Al/^{27}Al ratio at the start time of a sandglass, which is called the initial ^{26}Al/^{27}Al ratio. If we compare the initial ^{26}Al/^{27}Al ratio of different planetary objects, we can infer which object was born earlier in the Solar System. An isotopic ratio associated with a short-lived isotope such as ^{26}Mg/^{27}Al can thus be used as a unique radiometric clock especially suited to dating an event that occurred in the early Solar System.

Lee, Papanastassiou, and Wasserburg applied this principle to a number of meteorites, and they found that a complex mineral known as CAI (standing for calcium–aluminum-rich inclusion), which widely occurs in unaltered primitive meteorites, has the highest initial ^{26}Al/^{27}Al ratio among planetary objects. That is, CAI represents the earliest born or the oldest mineral in the Solar System. For this reason, it is customary to use the CAI as the origin of time when discussing the evolution of the Solar System.[5]

NOTES

1. For the details of various methods in isotope geochronology, see [6, 7].
2. The original article is [8]. A popular account of Rutherford's work can be found in [9].
3. The energetic activity of these three scientists in the early days of plate tectonics revolution can be read in [10].
4. The Precambrian refers to the time before the Phanerozoic, i.e. before 540 million years ago. The Phanerozoic eon is divided into three eras: the Paleozoic (540 million years ago to 250 million years ago), Mesozoic (250 million years ago to 65 million years ago), and Cenozoic (65 million years ago to present). The Paleozoic era is further divided into six periods: Cambrian, Ordovician, Silurian, Devonian, Carboniferous, and Permian. The Mesozoic era is divided into three periods: Triassic, Jurassic, and Cretaceous. The Cenozoic era is divided into two periods: Paleogene and Neogene. The oldest period in the Phanerozoic is the Cambrian, and this is why the time before the Phanerozoic is called the Precambrian.
5. For recent reviews on early Solar System history, see [11–13].

5   Plate tectonics revolution

PALEOMAGNETISM AND APPARENT POLAR WANDER

The Earth's magnetic field can be thought of as generated by an enormous magnet placed at the center of the planet. The points that pass through the Earth's surface when both poles of this enormous magnet are extended are called the north and south magnetic poles. The present north and south magnetic poles differ a little from those of the geographic poles, the difference being about 10° at present. However, if we take an average of the geomagnetic field direction over a few thousand years, the averaged magnetic poles almost perfectly coincide with the geographic poles. As mentioned in Chapter 3, the origin of the geomagnetic field is the fluid motion in the liquid outer core, and it is expected from the geodynamo theory that the magnetic poles align with the rotation axis of the Earth.

When magma erupts on the surface and cools down, it acquires magnetization in the direction of the ambient magnetic field, i.e. the geomagnetic field. The magnetization thus acquired is called thermo-remanent magnetization and has been shown to be extremely stable (*cf.* Chapter 9). Therefore, the remanent magnetization of volcanic rocks provides a very faithful record of the geomagnetic field at the time of eruption. Accordingly, by measuring the remanent magnetization of volcanic rocks with various ages, we can infer the direction of the geomagnetic field in the past.

If a rock is moved after it acquires remanent magnetization, we may infer its movement with reference to the magnetic pole, the information of which is imprinted on the rock. This is called paleomagnetism and is the most powerful tool in reconstructing the history of surface movements. For example, if we find a rock vertically magnetized on the equator, we suspect that the rock was originally formed at the north or south pole (depending on the direction of magnetization and the polarity of the geomagnetic field at the time of eruption) and was moved later to the equator by some means. When we measure a number of rocks with similar eruption ages from the same continent, however, we typically find that all samples suggest similar movements. Does this mean that the entire continent moved? It is not necessarily so. An alternative explanation is that the magnetic poles deviated considerably from the rotation axis of the Earth around that time; such a possibility is difficult to exclude purely on a theoretical basis. Assuming that a continent does not move, therefore, we can estimate the movement of the paleomagnetic poles by measuring the magnetization of rocks with a range of eruption ages. Such movement of the paleomagnetic poles is known as an "apparent polar wander" path. It is called "apparent" because the path can be interpreted as either the actual path of the magnetic pole movement or the path of the continental motion. With samples from one continent alone, we cannot distinguish between these two possibilities. We will see next how looking at samples from different continents resolved this ambiguity and provided a convincing answer to a long-standing problem of continental drift.

CONTINENTAL DRIFT HYPOTHESIS

On the left-hand side of Figure 5.1 is the present world map, showing the South American and African continents. On the right is a map in which these two continents are joined together. They fit together extremely well; Africa's Ivory Coast is completely wrapped around the tip of Brazil. This map was published in the book *La Création et ses Mystères Dévoilés* by a French geographer, Antonio Snider-Pellegrini, in 1858. Was this remarkable union of the two continents no more than mere coincidence? Snider-Pellegrini realized that similar fossils common to the American and African continents appeared in layers including coal, and began to suspect that the continents might have been joined together in the Carboniferous period when the coal was formed.[1] Figure 5.1 was his answer to this suspicion. He was not the first to propose that the continents may have moved, as the idea appears in

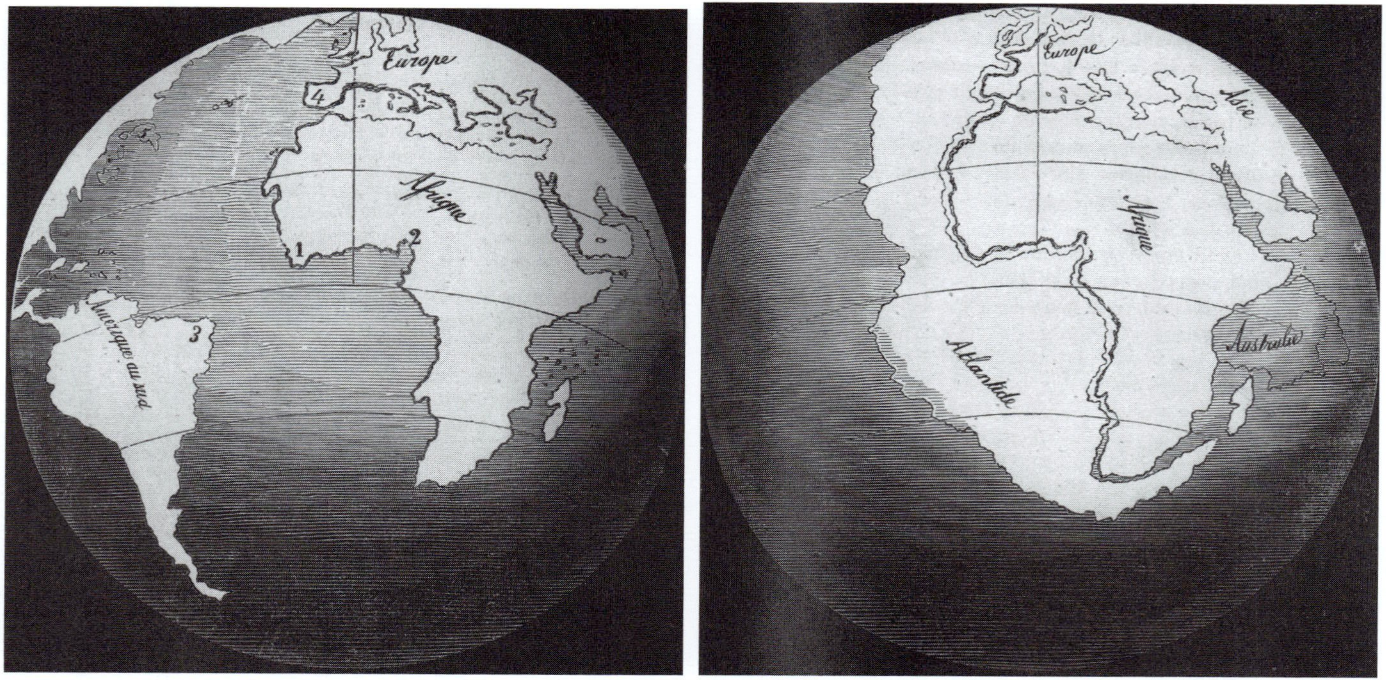

FIGURE 5.1 Snider-Pellegrini's conception of continental drift.[1]

the works of several pioneers such as Roger Bacon. However, Snider-Pellegrini's book added a new weight to the continental drift hypothesis by noting that fossils appearing on both continents were similar. A few decades later Alfred Wegener of Germany also became interested in this idea, and by assembling a variety of geological observations available at that time, he put forward a first comprehensive synthesis on continental drift in 1915.[2] Because it was such a radical idea at that time, however, Wegener's work was largely ignored as sheer speculation. As explained later, he was unable to answer satisfactorily a few critical questions on how continental drift works, and many felt that his arguments were weak, supported only by circumstantial evidence. It was not until the 1950s that scientists became certain that continental drift did occur. This turning point was brought about by the development of the paleomagnetic method.

At the beginning of the 1950s, Keith Runcorn and his colleagues in the United Kingdom collected rocks of various ages from around the world and measured their remanent magnetism, in order to construct apparent polar wander paths for different continents.[3] Figures 5.2a and b show such polar wander paths for South America and Africa. The corresponding ages are labeled on these paths. The difference between the location of the past and present poles steadily increases with age. More interestingly, the polar wander path found by using rock samples from South America is considerably different from that found by using African rocks. How did this difference arise? One can hypothesize the migration of the magnetic poles to explain the polar wander path for South America, but the same migration cannot be applied to that of Africa. Runcorn and his colleagues concluded that the difference is due to the relative movement between the American and African continents. Let us actually try arbitrarily moving one continent so that these polar wander paths overlap each other. The results are shown in Figure 5.2c. When the tip of Brazil has been completely surrounded by the sweep of the Ivory Coast, the two polar wander paths match more or less perfectly. The similarity of this figure to that drawn by Snider-Pellegrini about a century ago is unmistakable.

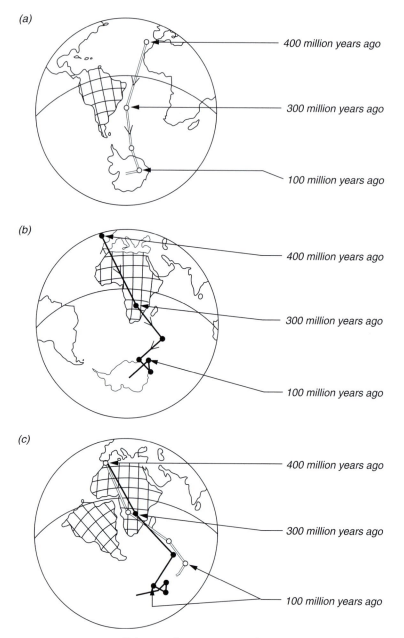

FIGURE 5.2 Polar wandering curve and continental drift.[4] (a) Polar wandering curve as determined from samples from South America. (b) Polar wandering curve as determined from samples from Africa. (c) The above two polar wandering curves almost coincide if South America and Africa are joined together.

Looking at the polar wander paths in Figure 5.2c in further detail, we can see that the two paths coincide up to about 200 million years ago, and from then to the present move away from each other in opposite directions. When the relative movements of the continents are taken into account, this can be understood perfectly. When the American and African continents were joined together as in Figure 5.2c, they traced the same polar wander path, and when both continents began to separate from each other, their polar wander paths started to deviate. The age at which the two polar wander paths began to deviate corresponds to the age at which the continents started to move apart. As seen in Figure 5.2c, this was about 200 million years ago. The movements of continents indicated by the remanent magnetization of rocks is surprisingly similar to the conclusions reached formerly by Snider-Pellegrini and Wegener based on geological observations. This discovery from paleomagnetism invigorated interest in the continental drift hypothesis and opened a door to a revolution in Earth sciences.

SEAFLOOR SPREADING THEORY

One of the main criticisms of Wegener's hypothesis on continental drift was the difficulty of continents moving through ocean basins, which are made of hard rocks. Continental blocks do not float on the oceans; they are welded to the oceanic crust, which lies beneath the oceans. How can continents plow through such hard rocks? Wegener could not answer this question, but in the 1960s, it was solved elegantly by the seafloor spreading theory.[4] The answer is deceptively simple. Continents are not plowing through ocean basins; instead, new ocean basins have been created as continents move. Thus, there is no need to break hard rocks to begin with. When South America drifted away from Africa, for example, a gap was created between these two continents. The mantle below rose to fill the gap, and as a result, a new piece of the ocean floor was produced. As these continents moved further and further away, the Atlantic Ocean basin grew continuously, and it is still growing today as South America is slowly drifting away from Africa.

New ocean floor is thus created between continents, or in the middle of oceans, and it is elevated higher than older ocean floor. The place where new ocean floor is created is thus called a mid-ocean ridge, and the most representative examples of mid-ocean ridges are the Mid-Atlantic Ridge running from south to north along the center of the Atlantic Ocean, and the East Pacific Rise, which runs roughly parallel with the Pacific coastline of South America. Each of these mid-ocean ridges is a great mountain range, towering up a few kilometers from the deep ocean floor, and its length stretches over more than 10 000 km. No ranges on land can rival this scale. These massive mountain ranges exist about 2 to 3 kilometers below the sea surface, so we cannot visibly observe them from space. Light can penetrate water without being attenuated only for a few meters, so in order to study such deep-sea features, we need to use echo-sounding techniques or visit them with a submersible. The first direct observation of a mid-ocean ridge was carried out from 1971 to 1974 under the FAMOUS (French–American Mid-Ocean Undersea Study) program, using the French-made *Archimède* and the American-made *Alvin* submersibles, in addition to various geophysical observations. The results of these observations supported the mid-ocean ridge structure hypothesized by the seafloor spreading theory.[1]

According to the seafloor spreading theory, the mid-ocean ridge where new materials upwell is the youngest part, and the ocean floor should become progressively older symmetrically to the left and right. This age pattern has actually been confirmed by examining fossils in the sediments covering the sea floor. The older the sea floor is, the thicker the sedimentary cover becomes, with the shallowest part being the most recently accumulated. The age of the sea floor should be similar to the age of the fossils contained in the deepest part of the sediment. To recover sedimentary rock samples from the deep ocean basins, the Deep Sea Drilling Project (DSDP) was initiated in 1968 by the United States, and a systematic increase in the age of fossils moving away from the Mid-Atlantic Ridge was discovered, providing a convincing proof of the seafloor spreading theory.[2] The DSDP has since grown into an international endeavor, succeeded by the Ocean Drilling Program in 1983 and by the Integrated Ocean Drilling Program in 2003. Combined with the age estimation based on the geomagnetic polarity scale (Chapter 4), the ages of the world's ocean basins are now well understood (Figure 5.3). For example, the age of the rock near the East Pacific Rise is practically zero, and increases as one moves further away to the right and left. The oldest part of the Pacific Ocean is the region immediately to the right of the Japan trench and the Izu–Mariana trench, which are furthest from the East Pacific Rise. The ages of these regions are estimated to be from the Jurassic to the Cretaceous periods (i.e. about 100–200 million years ago).

PLATE TECTONICS THEORY

From the late 1960s to the 1970s, the seafloor spreading theory was further developed into a more quantitative theory called "plate tectonics".[5] The movements of continents and the formation of new ocean floor at mid-ocean ridges are best understood if the surface of the Earth is divided into a dozen or so "plates" (Figure 5.4), which are moving on their own. A mid-ocean ridge is where two plates are separated, and the addition of new materials from below makes these diverging plates larger with time. As the surface area of the Earth is constant, if some plates are growing, other plates must be shrinking. The destruction of plates takes place when two plates are approaching each other, and usually one of them descends under the other and sinks back to the mantle (Figure 5.5); such place is called a subduction zone. A different type of plate boundary occurs where two plates are neither departing from nor approaching each other; instead, they slide past each other, and this type of plate boundary is called a transform fault. The most famous of these is the San Andreas fault in California. This new view of Earth was strongly advocated first by Tuzo Wilson at the University of Toronto and promoted by such young scientists as Jason Morgan of the United States, Xavier Le Pichon of France, and Dan McKenzie of the United Kingdom.[3] The advent of plate tectonics is now regarded as the beginning of modern geology. Tuzo Wilson once

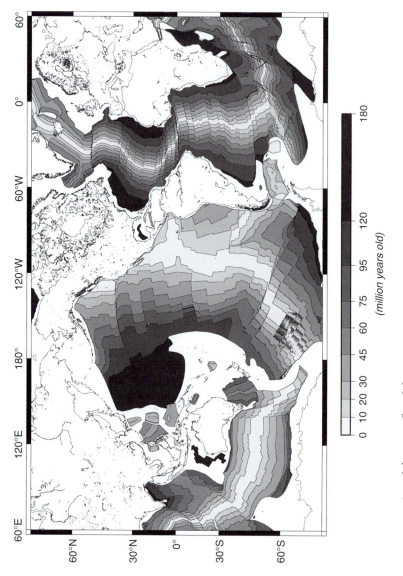

FIGURE 5.3 Age of the ocean floor.[5]

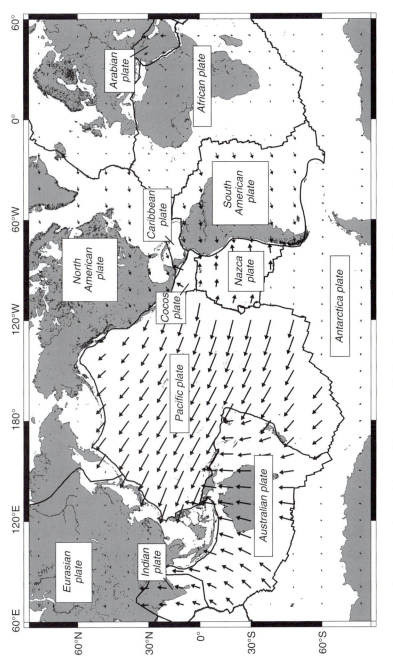

FIGURE 5.4 The tectonic plates of the world. Some small plates (e.g. Juan de Fuca, Philippine, and Scotia) are not shown here. Arrows denote present-day plate motion.[15] The longest arrow corresponds to about 10 cm/yr.

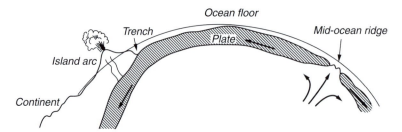

FIGURE 5.5 Schematic cross-section illustrating plate tectonic processes.

likened its arrival in Earth sciences to the discovery of quantum mechanics in physics in the 1920s.

The plate tectonics theory can explain a variety of geological observations in a unified framework. For example, as new sea floor, or a newly formed plate, moves away from a mid-ocean ridge and becomes older, it gradually cools down. Because of thermal contraction, a colder plate is denser and thus subsides. This is why mid-ocean ridges appear to stand out from the older ocean basins. Also, earthquakes almost always take place where two plates interact, e.g. rifted apart at a mid-ocean ridge or colliding at a subduction zone. The famous Pacific "Ring of Fire", where large numbers of earthquakes and volcanic eruptions occur, corresponds to subduction zones and transform faults. Using plate tectonics, we can also explain why mountains exist. When two approaching plates both carry continental blocks, these continents eventually collide, and this collision is what we see as mountain building. The formation of the majestic Himalayan Mountains is caused by the Indian sub-continent separating from Antarctica and Madagascar, all of which until then had been part of the same block, and moving north until it collided with the Eurasian continent. It is interesting that a true understanding of what builds mountains on land originates in the discovery of seafloor spreading under the seas. Collecting a variety of global data covering both continents and oceans was essential to arrive at the plate tectonics theory, and this partly explains why it was not possible to verify Wegener's hypothesis until the 1960s.

WHAT DRIVES PLATES?

The most critical weakness of Wegener's hypothesis is generally considered to be the lack of a driving mechanism. Wegener argued for continental drift by assembling a wide range of geological data, but he was not able to explain why continents had to move. Placed in the context of plate tectonics, this question can be rephrased as what drives plates. As soon as the plate tectonics theory appeared, geophysicists started to work on this problem, and we now know that plates move because of convection in the mantle. As mentioned in Chapter 1, the Earth is releasing heat from its surface to space, so the shallow part of the mantle becomes gradually colder and denser, and at some point, it becomes dense enough to start sinking. This is opposite to what happens when we boil water in a saucepan. In this case, water is heated from below by a stove, and the bottom water becomes hotter and more buoyant by thermal expansion. When the bottom water becomes sufficiently buoyant, it starts rising, leading to vigorous overturning in the saucepan. This is called thermal convection driven by basal heating, and what is happening in the Earth's mantle is instead thermal convection driven by surface cooling. Subducting plates correspond to sinking cold materials, and plate separation at mid-ocean ridges is caused by the pull of subducting plates. Put simply, plate tectonics is the surface manifestation of mantle convection, which is driven by planetary-scale cooling.

From the way seismic waves propagate through the mantle, however, it has long been known that the majority of the mantle is in the solid state. Only a very small fraction of the mantle can be in a partially molten state, so why can convection take place in the mantle? It was not until the 1950s that physicists discovered that even solid rocks could flow, albeit very slowly, by exploiting natural defects in the crystal lattice.[6] One way to measure the strength of a material is by viscosity.[7] For example, maple syrup is a hundred times more viscous than water, and peanut butter is a thousand times more viscous than maple syrup. The viscosity of the mantle is higher than that of peanut butter by about 20 orders of magnitude,[8] so it is extremely viscous and can move only very slowly. This is why we do not normally notice that the surface of the Earth is steadily deforming, except for rare occasions of large earthquake sliding. Viewed on the geological time scale, in which one million years is a typical unit, however, the mantle can be regarded as vigorously convecting, leading to substantial surface deformation such as plate tectonics.

NOTES

1. The Carboniferous period is in the Paleozoic era and extends from 360 million years ago to 300 million years ago. Extensive swamps and forests developed during this period, which later turned into coal beds. Carboniferous means "coal-bearing", and coal beds from this period provided much of the fuel for power generation during the Industrial Revolution.

2. Wegener's original work is in German,[7] but its English translation is widely available.[8]

3. Some representative studies on paleomagnetism from Runcorn's research group include [9, 10].

4. Seafloor spreading theory was proposed by [11], and its first scientific test using marine magnetic data was published by [12].

5. A large number of scientists contributed to the development of plate tectonics, and plate tectonics is now one of the core subjects in Earth sciences.[13–15] A good collection of original papers can be found in [16].

6. Theories of solid-state creep were developed in the field of materials science. [17–19] Though Earth's interior had long been believed to behave like a fluid over long time scales,[20, 21] its theoretical underpinning based on materials science was lacking until 1965.[22]

7. The viscosity of a material is proportional to the amount of stress required to deform the material at a certain rate. The SI unit of viscosity is Pa s, which may be understood as stress (Pa) per unit strain rate (1/s).

8. In scientific notation, "20 orders of magnitude" is expressed as 10^{20}. One million is 10^{6}, and one billion is 10^{9}. The representative viscosity of the Earth's mantle is about 10^{21} Pa s, whereas that of peanut butter is only 100 Pa s.

6 Evolution of the mantle

MANTLE EVOLUTION AND CRUSTAL GROWTH

Shortly after the Moon-forming giant impact, the segregation of the iron–nickel core from the primordial Earth was probably completed. The remaining fraction after the iron core separation is the mantle, but it was probably slightly different from the mantle we know at present, because another layer called the crust now exists on the mantle's outer surface (*cf.* Figure 3.1). As mentioned in the previous chapters, the crust is formed by the melting of the mantle. The mantle has been continuously supplying new crustal materials, and as a result, the chemical composition of the mantle itself has also been changing gradually. So the formation of the Earth's crust and the evolution of the mantle are complementary to each other; they are two sides of the same coin. As it is very difficult to obtain direct samples from the mantle, the evolution of the crust–mantle system has been studied mostly by looking at crustal rocks, which are much easier to sample.

The oldest rock in the crust is gneiss from a place called the Slave craton, which spreads over a substantial area in the northwestern part of Canada (a craton is a large, stable block of crust forming the nucleus of a continent). The age was measured by the uranium–lead dating method, which puts it at 4.03 billion years.[1] Rocks with ages slightly younger than 4 billion years are now found in several continental cratons including South Africa, Australia, and Greenland, and almost continuous occurrences of geological formations exist from then to the present.

The oldest minerals found in some rocks can go back even further. Some zircons found in sedimentary rocks from Jack Hills, western Australia, are dated to have formed about 4.4 billion years ago, again by the uranium–lead method.[2] Zircon ($ZrSiO_4$) is a tiny crystal generally less than a millimeter in size, and is thought to crystallize from existing crustal rocks. Because of its high resistance to weathering, zircon can survive even after host igneous rocks have been entirely eroded away. These ancient zircons either suggest that a substantial volume of the crust may already have existed on Earth as much as 4.4 billion years ago, or merely indicate an accessory mineral contained in some exceptional rocks.

It is not known whether the significant age difference (about 400 million years) between the oldest mineral and the oldest rock signifies that full-scale crustal formation did not occur until 4 billion years ago or is simply a matter of not finding rocks older than this. In this regard, an interesting speculation is that gigantic meteorite bombardment around 3.9 billion years ago, which is well recorded from numerous lunar craters and known as "lunar cataclysm", may have erased the earlier crustal records on Earth. Alternatively, we may speculate that the evolution of the mantle started to function fully to produce continental crust around 4 billion years ago. We will discuss this issue qualitatively on the basis of the samarium–neodymium isotope system.

NEODYMIUM ISOTOPE AS A MARKER FOR MANTLE EVOLUTION

Neodymium (Nd) belongs to the lanthanoid series, which in turn is part of the transition metals in the periodic table of elements. The lanthanoid series contains 15 elements, from lanthanum (La, atomic number 57) to lutetium (Lu, atomic number 71), which are usually all squeezed into one place in the periodic table because they are chemically very similar. As a side note, one row below from the lanthanoid series on the periodic table is the actinoid series, which contains thorium and uranium. Neodymium (atomic number 60) has seven stable isotopes, but one of them, neodymium-143 (^{143}Nd), is a decay product of samarium-147 (^{147}Sm), which has a half-life of about 100 billion years. Because of this very long half-life, changes in the amount of ^{143}Nd over the Earth's history are minute, and the geological application of this isotope system was not possible until the late 1970s when sufficiently accurate mass spectrometers became available.

Even though the signal of radioactive decay is so subtle, this pair of neodymium and samarium is particularly attractive as a marker for mantle evolution for the following reasons. Samarium (Sm, atomic number 62) is also a member of the lanthanoid series, so neodymium and samarium are chemically very similar to each other. Both elements do not like to be with iron, so when the core segregated from the primordial Earth, they stayed in the mantle side. But they are too big to fit within the crystal lattice of silicate minerals, and when the mantle melts, both of them readily escape to the melt phase. Because of their chemical similarity, therefore, they have almost always been together. Neodymium is, however, slightly larger than samarium, and it enters the melt phase slightly more easily than samarium. As a result, when new crustal materials are extracted from the mantle as magma, the leftover mantle has a little less neodymium than samarium. In other words, the Sm/Nd ratio of the mantle slightly increases after melting. As usual, we use the isotopic ratio when discussing the isotope evolution, and it is customary to use ^{144}Nd, which can be regarded as a stable isotope, as a denominator.[1] A higher Sm/Nd ratio in the mantle after crustal extraction means a higher $^{147}Sm/^{144}Nd$ as well, and owing to the radioactive decay of ^{147}Sm, it will lead to an increase in the $^{143}Nd/^{144}Nd$ ratio in the mantle.

Now consider what happens to the extracted crustal material. As mentioned, it is very difficult to sample the mantle directly, and we need to study crustal rocks instead to uncover what has happened in the mantle. When magma is generated in the mantle, more neodymium than samarium goes into the magma, but different neodymium isotopes should behave chemically identically, so the $^{143}Nd/^{144}Nd$ of the magma should be the same as that of the mantle at the time of eruption. Because the erupted magma contains samarium (including ^{147}Sm), the $^{143}Nd/^{144}Nd$ of the solidified magma (i.e. crustal rocks) gradually increases with time, but it is possible to backtrack the initial $^{143}Nd/^{144}Nd$ at the time of the eruption by finding the age of the rock using a radiometric dating method. By analyzing crustal rocks with various ages, therefore, we can infer how the $^{143}Nd/^{144}Nd$

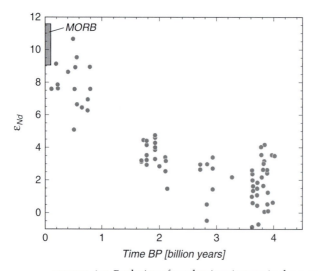

FIGURE 6.1 Evolution of neodymium isotope in the mantle,[3] as recorded by crustal rocks formed over the past four billion years (circles). The vertical axis indicates the deviation of $^{143}Nd/^{144}Nd$ from the (hypothetical) undifferentiated mantle. Higher ε_{Nd} corresponds to higher $^{143}Nd/^{144}Nd$. The shaded line (top left) denotes the range of present-day values found in mid-ocean ridge basalts (MORB).

ratio of the mantle has changed, which reflects the Sm/Nd ratio of the mantle.

As seen in Figure 6.1, the $^{143}Nd/^{144}Nd$ ratio of the mantle as estimated this way continuously increases over the Earth's history (though data are rather scattered), which is consistent with the idea that crustal materials have continuously been extracted from the mantle. Similar conclusions have been reached by studying other isotope systems, such as the lutetium–hafnium and rhenium–osmium systems.[3]

WHY AND HOW MANTLE MELTS

Mantle melting results in the growth of crust, but why does the mantle have to melt to begin with? To understand this, we need to realize that phase changes usually depend on pressure. Perhaps the most familiar phase changes are water freezing into ice, or water boiling to steam; the former is a change from a liquid phase to a solid phase, and the latter is from a liquid phase to a gas phase. At one atmospheric pressure, boiling is known to take place at 100 °C, but at a lower pressure, it happens at a lower temperature. For example, the boiling point of pure water is 96.8 °C at the altitude of 1 km (0.9 atmospheric pressure) and 93.3 °C at 2 km (0.81 atmospheric pressure). People living in mountain ranges are well aware of this fact because the low boiling point directly affects various cooking processes; for example, they have to cook longer than a regular recipe suggests. Conversely, water does not boil in the deep ocean even when it is heated to 400 °C because of high pressure.

The melting of the mantle is also a phase change, and the melting point increases at higher pressures. When the mantle rises beneath a mid-ocean ridge to fill a void created by diverging plates, mantle rocks are moved from a deep, high-pressure region to a shallow, low-pressure region. The temperature of rising mantle rocks eventually exceeds the melting temperature, and they start to melt.[2] This is how the mantle melts. Notice that no heating is applied to melt mantle rocks; simply bringing rocks from some depth results in melting. This is different from our daily experience of melting; you need to heat a saucepan to melt a chunk of butter in it. If the mantle is too cold, say colder than 1000 °C, it will not melt even when brought up to the surface because it is still below the melting point. The upwelling of mantle materials beneath a mid-ocean ridge always takes place in plate tectonics, but the creation of new oceanic crust at the ridge requires the mantle to be hot enough to cross the melting point on its way to the surface.

At a subduction zone, one plate is going down under the other plate, and this also causes the melting of the mantle, but for a different reason. Plates beneath oceans absorb a large amount of seawater, and when such a plate is descending in the mantle at a subduction zone, it releases the water into the surrounding mantle. From the viewpoint of mantle rocks, water is an impurity, and the presence of impurity lowers the melting point. Just as snow melts when you sprinkle salt grains over it, the mantle starts to melt when water is injected. This is why a chain of volcanoes forms at a subduction zone. As a subduction zone (and thus a volcanic chain) forms an arc on the globe, this type of magmatism is called arc magmatism, and the Pacific Ring of Fire is the most prominent example of it.

Mid-ocean ridge magmatism, which creates new oceanic crust, and arc magmatism, which creates volcanic arcs along subduction zones, are both caused by plate tectonics and take place at plate boundaries. There is another type of magmatism that is responsible for the formation of oceanic islands such as the Hawaiian islands. This is called hotspot magmatism. Because it often takes place in the middle of a plate, it is difficult to explain with plate tectonics. If the mantle melts by its vertical movement as in the case of mid-ocean ridge magmatism, why does the mantle have to move upward in the middle of nowhere? The origin of hotspot magmatism is widely debated,[3] but one popular idea is that such mantle upwelling originates in the core–mantle boundary. As mentioned in Chapter 5, convection in the mantle is largely driven by surface cooling, but the mantle is also heated from below because the core is hotter than the mantle. Thus, mantle materials just above the core are continuously heated by the core and occasionally become buoyant enough to rise. This upwelling from the lowermost mantle could eventually result in volcanic activities at the Earth's surface, irrespective of plate boundaries because where such upwelling is initiated at the bottom of the mantle has nothing to do with how plates move at the top of the mantle.

Mid-ocean ridge magmatism, arc magmatism, and hotspot magmatism are three major types of magmatism. Among them, mid-ocean ridge magmatism is volumetrically most significant, responsible for about 70 percent of the entire magma generated on Earth.[4] Because it takes place underwater, we hardly notice it, but the global network of mid-ocean ridges comprises the largest volcanic system, dwarfing subaerial volcanoes such as Mauna Loa at Hawaii.

OCEANIC CRUST AND CONTINENTAL CRUST

As briefly mentioned at the beginning of Chapter 3, there are two kinds of crust, oceanic crust and continental crust. Both are the product of mantle melting, but they have very different characteristics. Oceanic crust eventually returns to the deep mantle at subduction zones, and because of this recycling by plate tectonics, the age of oceanic crust is limited to about 200 million years old. Older oceanic crust has all been subducted. Even though the creation of oceanic crust is the most significant magmatism on Earth, therefore, chemical differentiation by mid-ocean ridge magmatism does not affect the composition of the mantle in the long run. The average composition of oceanic crust is similar to that of basalt.

Conversely, arc magmatism takes place on a plate overriding a subducting plate, so arc volcanoes do not subduct. This type of magmatism represents the permanent extraction of melt from the mantle and thus gradually modifies the average composition of the mantle. Continental crust grows largely by arc magmatism. Hotspot magmatism can also contribute to the growth of continental crust if it takes place in the middle of continents. Even if it takes place in the middle of oceans, oceanic islands may not entirely subduct, and some fraction of them could accrete to the continental crust. The average age of continental crust is about 2 billion years, one order of magnitude older than oceanic crust. The crustal growth discussed in the context of the neodymium isotope earlier in this chapter referred to the growth of continental crust. Oceanic crust does not grow with time, and the term "crust" when discussing mantle evolution usually means continental crust. The average composition of continental crust is similar to that of andesite, which is somewhere between granite (representing the upper portion of continental crust) and basalt (the lower portion). Melting of the mantle is known to produce only basalt, so making the continental crust requires some additional processes after extracting melt from the mantle. What those processes are is still widely debated.[4]

Having these two kinds of crust is unique to Earth, and it is generally considered to be a result of plate tectonics. Without subduction, for example, water would not be injected into the mantle; no arc magmatism would then take place, and thus no continental crust would be generated. The Earth is the only planet that exhibits plate tectonics, and the crust of other Earth-like planets such as Mars and

Venus is known to be similar to the Earth's oceanic crust; that is, these planets have no continents, though they have no oceans either!

INTERNAL STRUCTURE OF THE MANTLE

Just as the core is internally divided into the solid inner core and the liquid outer core, the mantle has its own structure, albeit more subtle. From studying how seismic waves propagate through the mantle, it has long been known that there are major seismic discontinuities at the depths of 410 km and 660 km (Figure 6.2). A seismic discontinuity is a boundary with a sudden increase in density and/or seismic velocities, which causes some extra reflections from otherwise smoothly propagating seismic waves. At the 660-km discontinuity, for example, the mantle density is estimated to jump by as much as 10 percent. The part of the mantle above 410 km is called the upper mantle, the part between

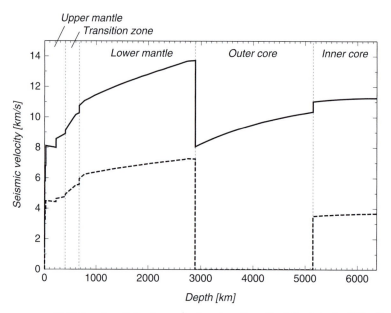

FIGURE 6.2 Seismic velocity in the Earth's interior.[10] Both compressional wave velocity (solid) and shear wave velocity (dashed) are shown. Shear waves do not propagate through a fluid, and the outer core is considered to be liquid because its shear wave velocity is zero.

410 km and 660 km is the transition zone, and the part below 660 km is the lower mantle. In some contexts, the upper mantle and the transition zone are collectively referred to as the upper mantle; that is, the mantle is divided into the upper and lower mantle by the 660 km discontinuity.

Why are there such discontinuities in the mantle? A common explanation is that they originate in the pressure-induced phase changes of mantle minerals. With increasing depth, pressure increases from 0.6 GPa at the top of the mantle to 136 GPa (about 1.4 million atmospheres) at the bottom of the mantle. With increasing pressure, one solid phase can change to another. An outstanding example is the transformation of graphite (the low-pressure form of carbon). If we compress graphite hard enough, it will become precious diamond (the high-pressure form of carbon). Diamond has a more compact crystalline structure than graphite, so the former is denser than the latter. A phase change with increasing pressure thus converts a material into its denser form.

In the late 1960s, the phase change of a common rock-forming mineral, olivine (olivine is believed to occupy about 60 percent of the shallow mantle), was suggested to be responsible for the mantle discontinuities, and the first decisive experiment was made by Ted Ringwood at the Australian National University.[5] He showed that olivine was transformed to a denser crystalline structure (later named wadsleyite) around a pressure corresponding to the 410-km discontinuity. This pioneering work triggered a new branch of Earth sciences called high-pressure mineral physics. Systematic work on the phase change of olivine has since shown that wadsleyite is converted to an even denser form of olivine (named ringwoodite in honor of Ted Ringwood) around a depth of 520 km, and ringwoodite is then decomposed into a combination of two minerals, (Mg,Fe)-perovskite and ferropericlase, at the depth of 660 km. Whereas the transition from wadsleyite to ringwoodite is known to take place over a range of pressure and does not produce a sharp seismic discontinuity, the olivine–wadsleyite transition and the breakdown of ringwoodite seem to take place at the correct pressures to match the 410-km and 660-km discontinuities.[6]

CHEMICAL STRATIFICATION IN THE MANTLE?

The above explanation of phase changes implies that the chemical composition of the upper mantle is the same as that of the lower mantle, even though the upper and lower mantle are made of different combinations of minerals. There are geochemical observations, however, suggesting that the mantle may be chemically stratified, and a number of geochemists have argued in the past that the mantle discontinuities cannot be explained by phase changes alone.[5]

As mentioned earlier, mid-ocean ridge magmatism produces oceanic crust, and hotspot magmatism creates oceanic islands. Both types of magmatism generate basalt, called mid-ocean ridge basalt (MORB) and ocean island basalt (OIB), respectively. MORB and OIB are known to have different isotope signatures. As an example, let us take a look at the helium isotope ratio $^3He/^4He$. Even among five rare gases, helium is the most inert element, so that it is impossible for it to form a chemical compound. Helium is incompatible with the solid phase, so when the mantle melts and magma erupts on the surface, helium is released from the solid Earth into the atmosphere. Because of its very light mass, helium eventually escapes into space in less than one million years.

Helium has two isotopes, ^3He and ^4He. Helium-4 has been continuously produced in the Earth's interior from the radioactive decay of uranium and thorium (^4He is the alpha particle emitted by alpha decay), but ^3He cannot be produced in a meaningful quantity within the Earth and is regarded to be primordial, i.e. inherited from the presolar nebula. In Figure 6.3, we compare the $^3He/^4He$ isotopic ratio (normalized by the atmospheric value) between MORB and OIB. For comparison, the $^3He/^4He$ ratios in the atmosphere and in continental rocks are also shown. The helium isotope signatures of MORB and OIB are different in two notable ways. First, MORB shows a much more limited variation than OIB in the $^3He/^4He$ ratio. Second, the majority of OIB samples have a higher $^3He/^4He$ ratio than MORB. What do these observations mean? Mid-ocean ridges span the globe, with a total length exceeding 60 000 km. It is thus surprising to see that rocks

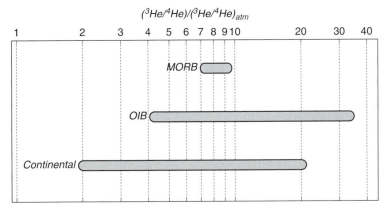

FIGURE 6.3 Helium isotopic ratio (^{3}He/^{4}He) observed in different types of rocks: MORB, OIB, and rocks from continental hotspots.[11] The isotopic ratio is normalized by the atmospheric value.

from various places along mid-ocean ridges have essentially an identical helium isotope signature.[6] One way to explain this remarkable isotopic homogeneity is to assume that the mantle is very well mixed by convection. This assumption does not require the entire mantle to be homogenized. Because we expect that only the shallow part of the mantle would rise in response to plate divergence at a mid-ocean ridge, only the upper mantle has to be well mixed. As mentioned earlier, the origin of oceanic islands is still controversial, but some of them may originate in the upwelling from the bottom of the mantle. So the isotopic diversity seen in OIB may be ascribed to the poorly mixed lower mantle. The higher ^{3}He/^{4}He ratio of OIB then means that the lower mantle is generally more primordial (i.e. less processed by mantle melting) than the upper mantle.

The contrast seen in the helium isotope signatures of MORB and OIB thus suggests a specific style of mantle convection; the upper mantle is vigorously convecting and efficiently mixed, whereas the lower mantle convects more slowly, leaving chemical heterogeneities relatively untouched. More importantly, mass transfer between the upper and lower mantle must have been severely limited throughout the Earth's history. Otherwise, we cannot maintain the different isotopic signatures of these two mantle layers. This style of convection is called layered-mantle convection, which is possible if, for example, the lower mantle is more iron-rich and thus compositionally denser than the upper mantle. Even though phase changes can adequately explain the presence of seismic discontinuities, therefore, the helium isotope data suggest that compositional differences may also contribute to these discontinuities. Similar conclusions can be reached by studying other isotope systems, and layered-mantle convection has long been preferred by geochemists.

STRUCTURE OF MANTLE CONVECTION

Our current understanding of the Earth's internal structure is primarily deduced from seismology, more specifically from analyzing elastic wave propagation initiated by earthquakes. The essence of this approach is to measure how long it takes for seismic waves emitted from an earthquake source to arrive at a number of observation sites distributed around the world. By analyzing a great number of earthquakes this way, seismologists can deduce the seismic velocity structure of the Earth's interior. Different kinds of silicate rocks have different seismic velocities, and even the same rock can have different velocities at different temperatures and pressures. The seismic velocity structure can thus provide extremely important constraints on the material properties and physical states of the Earth's interior. Owing to recent developments in computer technology, seismologists are now able to use not only the arrival time but also the shape of the seismic waves to estimate a more detailed velocity structure of Earth's interior.[7]

The technique seismologists use to reconstruct the three-dimensional seismic velocity structure of Earth is called seismic tomography, which is very similar to the computerized axial tomography used in medicine (known as CAT scans). In fact, seismic tomography and CAT scans were developed almost concurrently in the 1960s. Whereas the attenuation of X-rays is used to estimate the density structure of a human body in a CAT scan, the travel time of elastic waves is used to estimate the velocity structure of the Earth in seismic tomography. Seismic tomography allows us to see, for example, how deeply subducting plates sink through the mantle. Because subducting plates are colder than the surrounding mantle, and colder materials generally have faster seismic velocities, they are characterized by positive velocity perturbations. Results from high-resolution seismic tomography became available in the mid-1990s, and some of the positive velocity perturbations that originate from subduction zones have been shown to continue down to the bottom of the mantle (Figure 6.4). Seismology thus points to considerable mass transfer between the upper mantle and the lower mantle. Can this be reconciled with geochemical observations?

Since the late 1990s, debates on a unifying theory of mantle convection, which can explain both geophysical and geochemical observations, have intensified, and one lesson we have learned from them is that the connection between mantle convection and surface magmatism may not be as simple as previously thought. Consider the contrast seen in the helium isotopic signatures of MORB and OIB. Does this simply reflect the difference in mantle sources from which those basalts are generated? Or does it originate in different melting mechanisms? The way new oceanic crust is formed is surely very different from the way oceanic islands are created. Because a much larger volume of the mantle is involved in mid-ocean ridge magmatism than hotspot magmatism, the isotopic signature of MORB may appear to be averaged and more homogenized than that of OIB.[7] Also, the more primordial feature of OIB samples has been used to argue that the deep mantle has to be sequestered from convection, but how efficiently does mantle convection actually homogenize the mantle? With the current rate of mantle melting at mid-ocean ridges, it is known to take about 5 billion years to process the entire mantle.[8] Was the mantle processed more quickly or more slowly in the past? Trying to understand the meaning of geochemical observations thus requires us to better understand how the Earth has evolved over the past 4.6 billion years. Distinguishing observations and interpretations is not always easy but is essential. It is important to critically evaluate

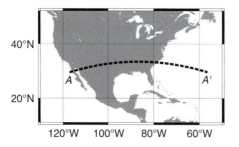

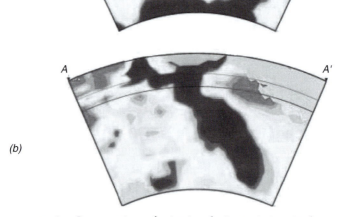

(a)

(b)

FIGURE 6.4 Cross-sections of seismic velocity variations in the mantle along a section through the southern United States.[12] (a) Compressional wave velocity; (b) shear wave velocity. Dark and light shades indicate positive and negative deviations from the reference, respectively. The positive velocity variation extending across almost the entire mantle in these figures is considered to reflect a subducted plate.

different kinds of data from geophysics, geochemistry, and geology and synthesize them in the context of planetary evolution, and we have just begun to appreciate the significance of such multidisciplinary endeavor.[9]

NOTES

1. Neodymium-144 is radioactive, but its half-life is 2.3×10^{15} years, which is 5 million times longer than the age of the Earth, so it can be regarded as a stable isotope for geochronology.
2. The temperature of rising mantle rocks actually decreases because of adiabatic decompression. But the drop in the melting point is steeper, so the rising rocks eventually become hotter with respect to the melting point. The discipline that studies the melting of the mantle is called igneous petrology.[13, 14]
3. http://www.mantleplumes.org/
4. The genesis of continental crust over the Earth's history is a challenging problem because of the complexity of chemical and physical processes involved and because of the possibility of secular changes in such processes.[15–18]
5. Some representative papers include [19–21].
6. This homogeneity of MORB isotopic signatures, however, is partly owing to how MORB is defined. For example, Iceland lies on a mid-ocean ridge, and Icelandic samples show highly primordial helium signatures, but they are classified as OIB.
7. Seismology is to Earth sciences what the eye is to the body. For introductory materials, see [22, 23].

7　Origin of the atmosphere and oceans

The seemingly simple conclusion that the Earth was not born in a space filled with air, but that the air was formed after the birth of the Earth, is the starting point in considering the origin of our atmosphere. This was pointed out clearly for the first time in 1947 by Harrison Brown of Caltech.[1] When considering the origin of the Earth's atmosphere, Brown turned his attention to the abundance of rare gases in the atmosphere. There are five types of rare gases. In order from the lightest to the heaviest atomic weight, they are helium, neon, argon, krypton, and xenon. Since none has any chemical affinity, they hardly ever combine with other atoms and are extremely inactive. Because of this, a rare gas is also called a noble gas (i.e. too "noble" to bind with other elements), but the term "rare gas" actually hints at the origin of the atmosphere. As this term implies, the atmosphere contains only an extremely small amount of these (argon-40 is the only exception, which will be discussed later). In terms of volume, xenon constitutes no more than about 0.000 01 percent, or 0.1 ppm (parts per million), of the atmosphere, and even neon, which is the most abundant rare gas (excluding argon-40), constitutes only about 20 ppm.

Looking at the whole Solar System, however, rare gases are by no means rare, compared with other elements with similar atomic weights. Recall the solar abundance of chemical elements discussed in Chapter 2; neon (atomic number 10) is as abundant as nitrogen (atomic number 7) and magnesium (atomic number 12). The peculiar rarity of rare gases in Earth's atmosphere can be seen by plotting their relative abundances with respect to their abundances in the Solar System (Figure 7.1). For example, neon in the Earth's atmosphere is depleted by ten orders of magnitude with respect to its abundance in the

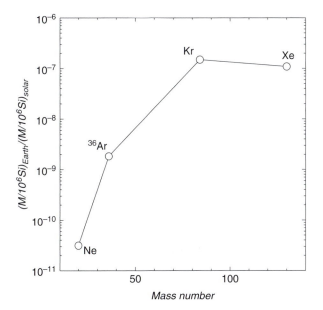

FIGURE 7.1 Abundance of rare gases in Earth's atmosphere with respect to solar abundance.[17] Both abundances are normalized with respect to 10^6 Si (Si abundance is set to 10^6 by convention; see Figure 2.3). For argon, only ^{36}Ar is considered here while other elements include all of their isotopes. Note that helium is not included here because it is extremely light; thus it easily shakes off the Earth's gravitation and escapes to space.

Solar System. Why are rare gases so rare in Earth's atmosphere? What does this rarity tell us? These are the questions raised by Brown in his 1947 paper.

Let us suppose for now that the presolar nebular gas was trapped intact as the Earth's atmosphere when the Earth was born. The composition of the atmosphere formed in this way would naturally be identical to the composition of the presolar nebular gas, so the abundance of rare gases within this atmosphere would also differ little from their abundance in the Solar System. However, Figure 7.1 shows a situation completely contrary to these expectations. Clearly, the Earth's atmosphere cannot be linked directly to the presolar nebular gas, and Brown suggested that it was formed after the birth of the Earth as something characteristic to the Earth. It is now believed that any primary atmosphere inherited from the presolar nebula was practically removed by an intense solar wind in the early Solar System.[2]

DEGASSING FROM THE EARTH

As Brown's research shows, the Earth's atmosphere must have been formed after the planet's birth. How exactly, then, was the atmosphere formed? Where did it come from and when? The first important clue was given in 1937 by the distinguished German nuclear physicist, Carl Friedrich von Weizsäcker.[3] In order to understand his idea, let us take another look at Figure 7.1. This figure deals with only one of argon's isotopes, argon-36, though with the other rare gases all of their isotopes are included.

Argon in the atmosphere is composed of the three isotopes, argon-36, -38 and -40, with argon-40 accounting for 99.6 percent of all argon. Although argon is said to be a "rare gas", it is the third most common element in our atmosphere after nitrogen and oxygen, accounting for about 1 percent of the volume of the atmosphere. Von Weizsäcker noticed this unusually large amount of argon-40 in the atmosphere and reasoned that it was probably formed through the nuclear disintegration of the potassium contained in the crust and mantle. In other words, he hinted at the idea of gas escaping from the solid Earth. About ten years later, it was confirmed that potassium-40 does indeed decay to argon-40 by electron capture.[4] Now we can explain why only argon-36 is shown in Figure 7.1. Because argon-40, which accounts for most argon, is formed by nuclear disintegration in the Earth's interior, it cannot be treated in the same way as other rare gases that existed when the Earth was born.

The other components of our atmosphere can also be explained in terms of volatiles that have escaped from the Earth's interior. So the origin of the atmosphere is basically about how volatiles have escaped from the Earth. We will thus discuss the degassing of the Earth next.

IS THE DEGASSING CONTINUOUS?

The first person to attempt a clear answer to the question of how volatiles escaped from the Earth was William Rubey of the US Geological Survey. Rubey first examined weathering of crustal rocks as the most likely way for volatiles to degas. Crustal rocks contain potentially volatile components that can become part of the atmosphere, for example nitrogen and water. When rocks weather and crumble into pieces, some of these volatile components may be released into the atmosphere. Rubey consequently tried to estimate their abundance on the Earth. According to his estimates (and excluding sedimentary rocks), they account for less than 1 percent of the volatile components contained in the present atmosphere and oceans and in sedimentary rocks. The volatile components in sedimentary rocks originally existed in the atmosphere and oceans in the form of deposits of calcite, so in a broad sense, they should be regarded as part of the atmosphere and oceans. Based on this estimate, Rubey concluded that the remaining 99 percent or more must have been taken into the atmosphere and oceans directly from the mantle. Rubey published this conclusion in 1951, and it is still considered to be useful.[5]

Rubey also considered the form of the volatiles that escaped from the mantle, suggesting that they escaped as volcanic gas and hot spring waters. He estimated the amount of gas released from the world's volcanoes (including volcanoes on the ocean floor) and the amount of hot spring waters released. If we suppose that gas has been continuously released at the present rate over the past 3 billion years, the total would be more than a hundred times the volume of the present Earth's atmosphere and oceans. It is known, however, that the volcanic gas (nearly all of it is water vapor) and hot spring waters are merely surface water that soaked down underground and was then recycled to the surface. If about 1 percent of this was drawn directly from the mantle, a sufficient amount would have been supplied from the Earth's interior (mantle) over 3 billion years to form the present atmosphere and oceans. This was the conclusion reached by Rubey.

Rubey's theory of Earth degassing (called the continuous degassing theory) has appeared in many textbooks and can be regarded as a classical view on the origin of the atmosphere and oceans. Later studies, however, indicate that this theory has various drawbacks. For example, metamorphic rocks of sedimentary origin make up a considerable portion of the world's oldest rocks, and it is clear that oceans on a scale sufficient to form these sedimentary rocks were already in existence 3.7 billion years ago.[6] There is little essential difference between the chemical composition of Precambrian sedimentary rocks and that of much younger sedimentary rocks, and this indicates that the nature of the Precambrian oceans was hardly any different from that of later oceans. If oceans of virtually the same scale as present were already in existence in the Precambrian, then we also expect the presence of an atmosphere with a similar chemical composition to that of the present atmosphere, because the atmosphere and oceans are always in chemical equilibrium on a geological time scale. This runs counter to Rubey's idea of continuous growth, i.e. the gradual development of the oceans and atmosphere throughout the Earth's history. Rubey's work is notable for its pioneering nature, but we should also note that many of his arguments were rather circumstantial. In order to proceed further, it is important to base our discussion on quantitative data, instead of circumstantial evidence. We will now explain one approach based on the isotopic ratios of rare gases.

USING THE ARGON ISOTOPIC RATIO

Let us first look at Table 7.1, which shows the relative proportions of the three argon isotopes in the present atmosphere. The amount of argon-40 is gradually increasing on the Earth owing to the disintegration of potassium-40. The isotopic ratio of argon-40 to argon-36 ($^{40}Ar/^{36}Ar$) in the present atmosphere has a value of 295.5. What was the ratio when the Solar System was born, 4.6 billion years ago?

In the case of relatively light elements, the smaller the mass number of the isotope, the more likely the isotope will be formed. According to nucleosynthesis theory, the ratio of $^{40}Ar/^{36}Ar$ when the nucleosynthesis ended would have been about one ten-thousandth. This means that when the Solar System was born, the argon present

Table 7.1 *Argon isotopic composition in the atmosphere (percent atomic abundance).[17]*

Argon-36 (^{36}Ar)	0.337
Argon-38 (^{38}Ar)	0.063
Argon-40 (^{40}Ar)	99.6

was nearly all argon-36 and argon-38. Naturally, some rare gases, including argon, would have been trapped in the interior of the Earth when the Earth was formed. Soon after the Earth was born, then, only a small amount of the argon in its interior would have been of potassium-40 origin, and the value of the argon isotopic ratio would still have been roughly one ten-thousandth. However, as time passed this value would have increased considerably, as illustrated in Figure 7.2.

If gas escaped from the Earth's interior in the extremely early stages of the Earth's evolution, the argon isotopic ratio at that time would not have been much greater than one ten-thousandth, and this does not correspond with the value of 295.5 in the present atmosphere. On the other hand, if the gas escaped extremely recently, the value of the argon isotopic ratio would be far greater than 295.5. These explanations assume that the gas escaped suddenly from the Earth's interior at a specific time, but even if the gas escaped continuously, the value of the isotopic ratio of the argon released into the atmosphere can be calculated. What is important here is that the value of the isotopic ratio of the argon in the present atmosphere is closely related to how gas was released from Earth's interior. Did it occur suddenly or was it continuous? If it occurred suddenly, when did it happen? If instead the degassing was more gradual, how slowly did it occur? Different ways of degassing correspond to different isotopic ratios of argon, so the present-day ratio of 295.5 can tell us how the atmosphere was formed in the past.

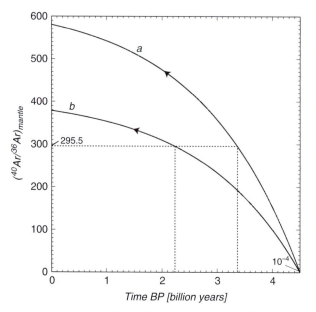

FIGURE 7.2 Examples of the isotopic evolution of argon in the mantle. Curves *a* and *b* correspond to the potassium concentration of 150 ppm and 230 ppm, respectively, in the mantle. Here ^{40}Ar produced by the decay of ^{40}K is assumed to be trapped in the mantle over the Earth's history. The value of ^{40}Ar/^{36}Ar of 295.5 corresponds to the present-day atmospheric value.

FORMATION OF THE ATMOSPHERE OCCURRED SUDDENLY

When we try to use the argon isotopic ratio to find out how gas actually did escape from the Earth, however, various difficulties arise. In order to draw the curves shown in Figure 7.2, it is necessary to know the amount of potassium in the Earth's interior and the amount of argon trapped in the interior when the Earth was born, although in practice it is extremely difficult to estimate these values accurately. Though the value of the isotopic ratio of the argon in the atmosphere is extremely important when considering the origin of the atmosphere, this alone is insufficient to elucidate the Earth's degassing, and other constraints are necessary. For example, the isotopic ratios of the argon *in the Earth's interior* are extremely useful. These ratios, if available, can considerably narrow the range of possible ways of degassing. Since a rather lengthy numerical calculation is involved, we will omit inter-mediate steps and describe results only.[1]

From the results of various experiments, the average value at present of the isotopic ratios of the argon in the Earth's interior (or more precisely, the mantle) is estimated to be at least 5000. Calculations show that, in order to satisfy this value, degassing must have occurred virtually instantaneously at a fairly early stage in the Earth's history (less than 500 million years after its birth). It can also be shown that this early degassing must have been quite intense, releasing more than 80 percent of the argon in the solid Earth. Hence the difference between the isotopic ratios of the argon within the Earth and that in the atmosphere cannot possibly be explained by the continuous degassing theory. After the early catastrophic degassing, however, the argon produced from the decay of potassium-40 in the mantle has been degassing continuously by volcanic activities throughout the history of the Earth, so we may say that the continuous degassing theory is partially correct.

We have focused on argon, but similar results can be expected for the other components of the atmosphere and oceans, e.g. nitrogen and water. In order to release argon from a rock in the laboratory, the rock must be heated to almost its melting point. From this laboratory result, it would be appropriate to consider that in a process violent enough to cause more than 80 percent of argon in the solid Earth to be released instantaneously, a certain amount of other volatile elements would also degas (though not to the same extent as argon). This conclusion of early catastrophic degassing is consistent with the existence of oceans in the early Earth, and also with the notion of the Moon-forming giant impact, which would have put the entire Earth into a magma ocean state.

XENON-129: A MORE PRECISE CHRONOMETER
FOR MANTLE DEGASSING

The very early degassing of argon from the mantle can be more quanti-tatively concluded from the comparison of xenon isotopes between the atmosphere and the mantle. In Chapter 2 we noted that some of the xenon-129 in the Earth was produced in the mantle through the radioactive decay of iodine-129, and subsequently degassed into the atmosphere together with other rare gases and volatiles. Let us also remember that the half-life of iodine-129 is only 16 million years, much shorter than the age of the Earth. Therefore, after only a few hundred million years from the birth of the Earth, virtually all of iodine-129 had decayed to xenon-129 in the mantle, and thereafter there would be no change in the xenon isotopic composition in the mantle. If the atmosphere were derived from the mantle after then, we would see the same xenon isotopic composition in the atmosphere as in the mantle. On the other hand, if xenon degassing occurred shortly after the birth of the Earth, there would still have been iodine-129 in the mantle, and the xenon in the present mantle would have more xenon-129 relative to the xenon in the atmosphere.

The excess xenon-129 in the mantle-derived rocks relative to the atmospheric xenon has been confirmed, and the very early rare gas degassing from the mantle is now difficult to refute.[7] Researchers have been trying to better quantify the degree of the excess xenon-129 in the mantle, since it would resolve the degassing time more precisely than argon isotopes can because of the much shorter half-life involved.

RARE GASES IN MANTLE MATERIALS

As seen above, considering the argon isotopic ratio opened a way to quantitative discussion on the origin of the Earth's atmosphere. Estimating this ratio for the mantle is, however, difficult because argon in the atmosphere can be mixed into rock samples you want to measure. Normally the amounts of argon and the other rare gases contained in volcanic rocks are extremely small. The argon contained in a gram of volcanic rock is no more than a hundred-millionth of a cubic centimeter. On the other hand, the atmosphere contains about 1 percent argon. If even a tiny amount of air is mixed with a rock sample, therefore, the argon from the air will far exceed the argon in the rock, and so it becomes difficult to ascertain the isotopic ratio of argon in the rock.

In order to find this isotopic ratio, various materials of mantle origin have been examined, including diamonds. It is thought that diamonds are formed at a depth of more than 100 km into the Earth's interior. It is virtually certain that they are of mantle origin, though there are some exceptional samples that may have been formed at a comparatively shallow depth by meteorite impact shock. Consequently, if diamonds contain rare gases, these rare gases can be considered to be of mantle origin in most cases. Diamonds have the excellent advantage of not being affected by any kind of chemical matter at all, and are extremely stable under high temperature. In addition, the crystals are extremely closely packed, so there is relatively little fear of gas trapped inside them escaping, or, conversely, of air and water from outside entering them. As a result, if rare gases are contained within the diamonds, they should reflect quite faithfully the composition of rare gases of mantle origin. The first successful experiment was done more than three decades ago by Takaoka and Ozima, who found an anomalous helium isotopic ratio in diamonds from Kimberley Mines of South Africa.[8] Owing to the enormous technical difficulties, only a few laboratories are currently engaged in investigating rare gases in diamonds, but future research will be vital given its high potential for resolving the indigenous rare gas isotopic composition in the mantle.

THE ORIGIN OF LIFE AND THE ATMOSPHERE

In terms of quantity, the atmosphere and oceans probably achieved levels similar to those in the present day in the very early stages of the Earth's evolution. But the composition of the atmosphere was likely to be vastly different. Oxygen, which accounts for one-quarter of the present atmosphere, was formed through the activities of life and would have been practically non-existent before life appeared on Earth. So what was the composition of the atmosphere when degassing had just occurred?

Some suggestions about the composition of the early atmosphere were made in the early 1950s by scientists interested in the origin of life, for example Harold Urey, a Nobel laureate at the University of Chicago. Urey and his colleague Stanley Miller attempted to create organic matter in the form of amino acids, a key ingredient for life, from inorganic matter. They discharged electricity into a mixed gaseous body of methane, ammonia, and water, and confirmed that organic matter was synthesized.[9] This is the famous Miller–Urey experiment. The reason Miller and Urey chose this seemingly strange gas composition for their experiment was as follows. Let us suppose that the composition of the atmosphere at the birth of Earth was similar to that of the presolar nebular gas, so that hydrogen, oxygen, nitrogen, and carbon possessed similar relative abundances to the presolar nebular gas in the early atmosphere. These elements react with each other to form stable chemical compounds. What chemical compound actually forms would be determined by the temperature of this early atmosphere. Miller and Urey adopted the general opinion of that time that the Earth accreted at a comparatively low temperature, and assumed a temperature of 25 °C. Then, it can easily be shown thermodynamically that the early atmosphere would have been a mixture of methane, ammonia, and water, and so this is what Miller and Urey used in their experiment. Lightning occurred in this early atmosphere, and organic matter was formed through electric discharge. Miller and Urey argued that this organic matter ultimately evolved into life. However, there are a few problems with this composition of the early atmosphere.

THE MYSTERY OF THE COMPOSITION OF THE EARLY ATMOSPHERE

The first problem is that Miller and Urey hypothesized that the elemental makeup of the early atmosphere was the same as that of the presolar nebula, but this hypothesis is unlikely to be correct. As already reasoned, the atmosphere was formed by volatiles degassing from Earth's interior. The composition of these volatiles should be different from the solar abundance (Figure 2.3). For example, there are extremely small quantities of rare gases in the Earth's atmosphere compared with those in the Solar System. In addition, observed volcanic gases commonly comprise water and carbon dioxide (generally amounting to more than 90%), but hardly contain any methane or carbon monoxide, which are the key starting molecules in the Miller–Urey experiment.[10]

The second problem was that the temperature of the early atmosphere was assumed to be 25 °C. In the case of degassing from Earth's interior, the temperature of volcanic gases would be naturally higher than 1000 °C. Thermodynamic calculations indicate that the atmosphere formed from such hot gases is composed of carbon dioxide, nitrogen, and water. It turns out that organic matter would not form by discharging electricity in this type of atmosphere.

Yet the Miller–Urey experiment represents by far the most plausible process for the production of organic matter from inorganic matter, bearing the potential to account for the origin of life on Earth. A number of attempts therefore have been made to resolve these difficulties.[2] One recent example, which seems promising, is to appeal to the specific physical and chemical conditions in the early Earth. Hashimoto of Kobe University and his colleagues realized that the putative magma ocean, which is supposed to have formed as a result of the Moon-forming giant impact, might hold an important key to the origin of life.[11]

A magma ocean state caused by a giant impact would have been a global version of the Kilauea lava lake, covering the entire surface of Earth to a depth of some hundreds to thousands of kilometers. As with the present volcanic magma, volatiles must have been efficiently degassed from the magma ocean in gigantic amounts, enough to yield the entire atmosphere (including oceans). The composition of degassing volatiles, however, was probably different. Some researchers believe that while the magma ocean degassed volatiles, iron metal was also being segregated from the magma ocean and fell to greater depths because of its higher density, so forming the iron core. In the case of degassing in the presence of iron metal, thermodynamic calculations suggest that the degassing volatiles essentially consist of hydrogen molecules, methane, carbon monoxide and dioxide, and nitrogen molecules. The resultant atmosphere is therefore close to the starting material employed in the Miller–Urey experiment, and we may circumvent the difficulty in relating the Miller–Urey experiment to the origin of organic matter in the early Earth.

BANDED IRON FORMATIONS AND PRODUCTION OF OXYGEN

To reiterate, it has been thought that the main components of the early atmosphere just after its formation by degassing were carbon dioxide, nitrogen, and water. However, recent studies suggest that the early atmosphere may have been composed of hydrogen, carbon monoxide, and methane, and such an atmosphere has the potential to produce organic materials, which are an indispensable ingredient for life. In either case, however, we do not have oxygen, which comprises about 20 percent of the present atmosphere. So when did the oxygen seen in the present atmosphere appear?

The appearance of oxygen is the most fundamental factor in considering the origin of life. Various proposals have been put forward by Earth scientists as well as biologists, and here we focus on the work of Preston Cloud, who first noted the connection between banded iron formations and the abundance of oxygen in the past atmosphere.[12] Banded iron formations are a type of sedimentary rock characteristic of the Precambrian period, and exhibit alternate layering of quartz microcrystals and iron oxide. Hardly any iron ore deposits with this remarkable combination of minerals are found after the Cambrian period (i.e. about 540 million years ago). Banded iron formations have also been found within the oldest rocks (approximately 3.7 billion years old) on the west coast of Greenland. So it is reasonable to think that the existence of these banded iron formations reflects the characteristic sedimentary environment in the Precambrian period.

To understand the implication of the existence of the banded iron formations, it is first necessary to know that there are two types of iron ions; ferrous ion has a +2 charge (Fe^{2+}) while ferric ion has a +3 charge (Fe^{3+}). Ferrous ion is easily soluble in water, but ferric ion forms insoluble products. In reducing environments – that is, in an oxygen-deficient atmosphere – iron will be in a ferrous state, which could be easily leached in solution, whereas in oxidizing environments iron is ferric and will form insoluble products such as iron oxide. From the existence of banded iron formations, therefore, Cloud conjectured that there must have been enough oxygen in the Precambrian oceans for iron oxide to form, sink, and deposit. On the other hand, for the iron to be able to be transported from the land to the oceans, it must have been leached from soil, which is only possible in an oxygen-poor atmosphere. Cloud concluded that the rhythmic banding could result from a fluctuating balance between oxygen-producing, living organisms in the ocean and the supply of dissolved iron from the land. The existence of the banded iron formations therefore suggests a very low concentration of oxygen in the atmosphere when the banded iron formations were deposited.

However, some researchers claim that, based on the carbon isotopic ratio of Precambrian calcium carbonates, they can trace the origin of oxygen in the ocean back to quite a long time ago (up to 3.3 billion years ago). This is the topic we will look into next.

APPEARANCE OF OXYGEN AND CARBON ISOTOPES

It is generally recognized that the existence of oxygen in the atmosphere is the result of the photosynthesis of plants. Various other processes by which oxygen could have been produced have been examined; for example, oxygen can be released by dissociating water vapor with the Sun's ultraviolet radiation.[13] But none of these theories can explain the large amount of oxygen in the present atmosphere. The presence of life appears to be fundamental for the formation of oxygen, as other planets without life, such as Mars and Venus, do not have oxygen in their atmospheres.

Based on this intrinsic connection between life and oxygen, some researchers suggested that the appearance of oxygen in the Earth's history might be timed by investigating the carbon isotopes of ancient rocks. Almost 99 percent of the carbon in nature has the mass number 12 (^{12}C), while the remainder has the mass number 13 (^{13}C). The isotopic ratio usually fluctuates by several percent depending on the environment. However, these fluctuations are quite regular. Carbon of organic origin contains a smaller proportion of ^{13}C than carbon of inorganic origin. This difference is quite clear, and so from the isotopic ratio of carbon, we can estimate whether a particular carbon is of organic or inorganic origin.

Let us now consider what happens with photosynthesis. Living plants take in carbon dioxide, and with the help of sunlight and chlorophyll, they release oxygen. Focusing only on the exchange of elements, photosynthesis is a process whereby one carbon atom (of inorganic origin) is taken into the body and changed into organic carbon, and one oxygen molecule is released outside the body. So one oxygen molecule corresponds to one organic carbon atom, and carbon will be more depleted in ^{13}C as more oxygen molecules are produced. Here we have to note that the physical and chemical process that gives rise to distinction between the inorganic and organic carbon is by no means fully understood. We come back to this issue in the next chapter.

Along this line of reasoning, Manfred Schidlowski and others tried to systematically investigate the isotopic ratio of carbon contained in Precambrian sedimentary rocks.[14] They measured samples from all around the world and discovered that they were clearly divided into one group with an ordinary amount of ^{13}C (inorganic origin) and another group with very little ^{13}C (organic origin). These two groups are clearly identified irrespective of sample age, and the existence of organic carbon can be traced to 3.5 billion years ago. Does this mean that the existence of life as well as atmospheric oxygen goes back to 3.5 billion years ago? Recent studies suggest that it may not be that simple.[15] Some suggest, for example, that the photosynthesis carried out by organisms in the early Earth was of a type that did not release oxygen, and if this is the case, the carbon isotope data only signify the presence of life but not oxygen. Others even suspect that some of the carbon isotopic variations reported by Schidlowski's group may be just due to a common isotopic fractionation known as mass-dependent fractionation, being irrelevant to photosynthesis. How can we test these possibilities? A revolutionary breakthrough came with sulfur isotopes.

SULFUR ISOTOPES, A PHOTOSYNTHESIS MARKER?

Leaving a technical explanation to the next chapter, we introduce one important recent discovery: sulfur isotopes are found to exhibit anomalous variations that cannot be attributed to ordinary mass-dependent isotopic fractionation, and more interestingly, there are good reasons to believe that these anomalies are related to the oxygen content in the atmosphere. Therefore, we can use this sulfur isotopic anomaly as a unique marker of oxygen in the atmosphere.

Figure 7.3 shows isotopic anomaly (defined as the deviation from ordinary mass-dependent fractionation) measured on a variety of rocks over the Earth's history, as first demonstrated by James Farquhar and his colleague at the University of California, San Diego.[16] It is remarkable that there are large variations in the isotopic anomaly before about 2.4 billion years ago. Also noteworthy is that the anomaly occurs in both a positive and negative sense from the mean value, in which positive anomalies characteristically occur in reduced sulfur (sulfides) and negative anomalies occur in oxidized sulfur (sulfates). The distinct association of the anomaly sign with the oxidation state of sulfur indicates that the anomaly is intrinsically related to the oxidizing condition in the atmosphere and oceans, and such fluctuating oxidizing environments cannot be produced with an oxygen-rich atmosphere. The absence

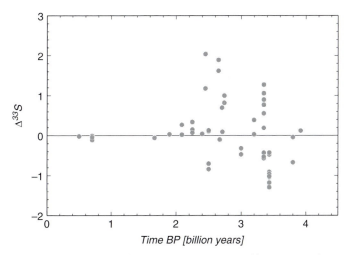

FIGURE 7.3 Sulfur isotopic variation in sulfur compounds in rocks with a variety of ages.[16] $\Delta^{33}S$ reflects the deviation of measured isotopic compositions from mass-dependent fractionation.

of a sulfur isotopic anomaly, therefore, is now regarded to be the most convincing time marker for the appearance of substantial oxygen in the atmosphere at around 2.4 billion years ago.

NOTES

1. For classical argon degassing models, see [18, 19]. Note that a recent experimental study suggests that argon may not easily enter basaltic magma,[20] potentially undermining the most fundamental assumption of these degassing models.
2. There are quite a few issues regarding the connection between the Miller–Urey experiment and the origin of life, and the composition of the early atmosphere is only one of them. For reviews on historical developments and current status, see [21, 22, 23]. There are also many popular books on the origin of life, e.g. [24, 25].

8   Isotopes as DNA of nature

NON-RADIOGENIC STABLE ISOTOPES

In the preceding sections, we have shown how isotopes play an important role in tackling a variety of problems concerned with tracing the Earth's evolution, with an emphasis on radioactive isotopes as a unique time marker. Let us recall, for example, that argon consists of three isotopes argon-36, argon-38, and argon-40, all of which are stable isotopes. Argon-40 is a radiogenic stable isotope, and its amount increases with time through the radioactive decay of potassium-40. The other two isotopes of argon are non-radiogenic stable isotopes, and their abundances have not changed since their birth in a star. We have discussed radiometric geochronology, in which the amount of a radiogenic stable isotope such as argon-40 yields an absolute time marker of rock or mineral formation age.

Here, we turn our attention to the use of non-radiogenic stable isotopes as a tracer of geochemical processes in nature. Thanks to the recent development of microchemical analysis of elements, systematic variations in the ratios of non-radiogenic stable isotopes, albeit extremely subtle, yield a unique means to entangle extremely complicated geochemical cycles of elements. Let us see how this becomes possible.

STABLE ISOTOPES AS COSMIC DNA

About 13 billion years ago, at the very beginning of the universe, the Big Bang produced primordial elements such as hydrogen, helium, and lithium from protons and neutrons. These primordial elements were then dispersed in an expanding universe. In Chapter 2, we briefly discussed how all elements (and isotopes) have been created from these primordial elements through nuclear reactions in stars. Elements have inherited their identity in the form of isotope signatures from the stars in which they were born. Elements ejected from stars via supernovae underwent gigantic mixing in galaxies and were finally isolated as a presolar nebula, from which planets including the Earth were formed. The isotopic characteristics of the Earth thus reflect those of this presolar nebula.

Once formed in a star, non-radiogenic stable isotopes remain immutable; they retain the original isotopic composition in accordance with specific nucleosynthesis processes in the star. The isotopic composition of an element, i.e. relative abundances of isotopes such as the ratio of argon-38 to argon-36, is uniquely determined by the physical characteristics (primarily mass) of a star and its chemical composition. Because of this immutability, stable isotopic ratios of an element can be used as a unique tracer, in an analogous way to DNA in biology. As an example, we will consider next the stable isotopes of oxygen, the third most abundant element in the universe.

OXYGEN ISOTOPIC COMPOSITION AND ITS VARIABILITY

Isotopes of an element are chemically identical, and therefore the isotopic composition is not affected by chemical reactions involving the element. For example, the isotopic composition of terrestrial oxygen should, in principle, retain the primordial isotope ratio inherited from the presolar nebula, irrespective of its chemical form such as oxygen molecule (O_2), carbon monoxide or dioxide (CO, CO_2), or water (H_2O), which are secondary products due to various chemical reactions.

Although chemistry by itself does not affect the isotopic composition of an element, the difference in the masses of isotopes generally affects the isotopic composition during physical and chemical processes. For example, when water evaporates from seas or lakes, water molecules containing the lighter oxygen isotope ^{16}O will more easily evaporate than those containing the heavier isotopes ^{17}O or ^{18}O. As the result, heavier isotopes will be enriched in water relative to vapor. By the same token, raindrops condensed in vapor cloud consist of heavier oxygen isotopes than their host vapor cloud.

Change in an isotopic composition, namely isotopic fractionation, is well explained by statistical thermodynamics.[1] The theory shows that the degree of isotopic fractionation is attributable to the difference of mass between corresponding isotopes. For example, the ratio of $^{18}O/^{16}O$ shows twice as much fractionation as that of $^{17}O/^{16}O$ whenever isotopic fractionation takes place in nature, in accordance with the mass difference of two for the former ratio relative to the mass difference of one for the latter.

From statistical thermodynamics we can also show that the total isotopic fractionation at chemical equilibrium can be expressed to a good approximation as a function of temperature and mass difference. Therefore, we can make use of this relation between isotopic fractionation and temperature to infer the temperature of an environment in which oxygen isotopes are equilibrated. The oxygen isotopic ratio $^{18}O/^{16}O$ has been most successfully used to infer environmental temperatures where oxygen existed in equilibrium between different phases such as water and vapor. The development of this field has owed much to the work by Harold Urey and his colleagues at the University of Chicago more than half a century ago.[2] With this technique, we are now able to trace the secular variation of atmospheric temperature back to tens of millions of years in the past.

PALEOTEMPERATURE AND CLIMATE

Most of Antarctica is now covered by thick ice, which has formed through re-crystallization of accumulated snow over many years. Reflecting the seasonal variation of snowfall rate, ice sheets generally show fine-scale layering, which can be used as relative time markers (as in the case of tree rings). Moreover, inclusions of volcanic ashes found in ice layers can often be identified to a specific volcanic eruption, thereby enabling us to assign an absolute age to the layer. To take advantage of the information contained in this layered ice, drilling of ice cores has been carried out in both Antarctica and Greenland, with some cores reaching depths of up to a few kilometers. Such cores cover a time span of hundreds of thousands years from the top to the bottom.

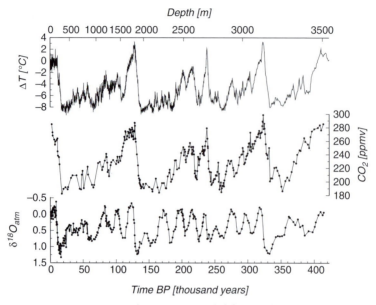

FIGURE 8.1 Vostok ice-core records.[7] Note that the paleotemperature reconstruction here uses hydrogen isotopes as well as oxygen.

An ice sheet contains tiny air bubbles, which were trapped from the atmosphere at the time of ice formation. By analyzing the oxygen isotopic ratio $^{18}O/^{16}O$ of the trapped air in bubbles, we are therefore able to infer the temperature in the past atmosphere. A striking example is shown in Figure 8.1. Based on the same principle, we can trace paleotemperature further back to hundreds of millions of years with the use of deep ocean-bottom sediments. In the abyssal ocean bottom far away from continents, the sedimentation rate is extremely slow, generally a few millimeters or less per thousand years. Therefore, a few hundred meters of sediments drilled from the ocean bottom would cover a time span of a few hundred million years.

Here we would like to comment briefly on global warming on the basis of paleotemperature estimates. As shown in Figure 8.1, the secular variation of atmospheric temperature, deduced from the isotopic ratio $^{18}O/^{16}O$ of air bubbles in the ice cores, almost perfectly coincides with that of the carbon dioxide content of the air bubbles from the present to

420 000 years ago. There seems little doubt that carbon dioxide content is intrinsically related to the atmospheric temperature and vice versa. An important question is, which was the cause of the relation? One may argue that a higher atmospheric temperature may have helped to release more carbon dioxide from the crust to the atmosphere. Alternatively, one may argue that a higher concentration of carbon dioxide in the atmosphere gave rise to a greater greenhouse effect to raise the atmospheric temperature. The greenhouse effect of carbon dioxide is undeniable, but how much of the temperature variation was actually caused by the greenhouse effect is still difficult to quantify, because the Earth's climate is a highly complex system with a number of fundamental processes yet to be understood.[1] Paleotemperature estimates over the geological time scale, however, are indispensable observational constraints on the dynamics of this complex system, continuing to provide new insights to the theoretical studies of climate.

MASS-INDEPENDENT ISOTOPIC FRACTIONATION

Although mass-dependent isotopic fractionation is common in natural processes, an entirely different type of isotopic fractionation, which is not related to mass, was discovered by Robert Clayton of the University of Chicago in tiny calcium–aluminum-rich mineral inclusions (CAI) in the Allende meteorite.[2] The oxygen isotopic composition of these inclusions does not follow the ordinary mass-dependent fractionation law predicted by classical statistical thermodynamics. Ten years after this discovery, Mark Thiemens and John Heidenreich at the University of California San Diego found that mass-independent oxygen isotopic fractionation also exists in the ozone layer.[3] However, it is still not clear if this fractionation is related to that in the CAI oxygen. Since its first discovery in 1973, this extraordinary isotopic fractionation has attracted the interest of a number of researchers, and quite a few theories have been proposed over the past four decades, but fractionation mechanisms are still widely debated.[4]

At present, more than a dozen elements are known to exhibit clear mass-independent isotopic fractionation in natural materials, including sulfur, calcium, and iron. In order to identify mass-independent isotopic fractionation, we need more than three stable isotopes so that we can see that isotopic fractionation deviates meaningfully from ordinary mass-dependent fractionation. Recall that on the basis of the carbon isotopic anomaly in Precambrian sedimentary rocks Schidlowski suggested the existence of atmospheric oxygen over the past 3.5 billion years (Chapter 7). Carbon has only two stable isotopes, however, and it was difficult to rule out the possibility that the carbon isotopic anomaly observed by Schidlowski was unrelated to mass-independent isotopic fractionation caused by photosynthesis, but instead caused by mass-dependent isotopic fractionation. Since sulfur has four stable isotopes, it is possible to discriminate mass-independent isotopic fractionation effects from ordinary mass-dependent fractionation. Because of this, we can be confident that the large scatters in the sulfur isotopic anomalies that occur before 2.4 billion years ago are caused by mass-independent isotopic fractionation reflecting the oxygen content in the atmosphere (*cf.* Figure 7.3).

The physical mechanisms responsible for mass-independent fractionation are still not well understood but are usually considered to be due to either non-equilibrium processes or quantum effects, or a combination of the two. Non-equilibrium processes are generally much more difficult to investigate than classical equilibrium cases, and studying quantum effects requires highly elaborate first-principles calculations. It is therefore understandable that the application of mass-independent fractionation has far preceded theory. The combined studies of mass-dependent and mass-independent isotopic fractionations are now flourishing in a variety of fields, especially in environmental sciences. As such, we briefly discuss the case of mercury (Hg).

MASS-INDEPENDENT ISOTOPIC FRACTIONATION
OF MERCURY

Mercury poisoning was the main factor in the tragedy of Minamata disease, which took place in the mid-1950s in a small country town called Minamata in the south of Japan. Industrial waste water from a chemical factory contained inorganic mercury, which itself is already toxic but was converted to even more toxic organic mercury compounds through the food chain by marine organisms. This accumulated in marine organisms such as shellfish, which were then eaten by the local population. As of early 2001, over 1700 deaths had been attributed to Minamata disease. Even after half a century, similar disasters are occasionally reported, and mercury is viewed as one of the worst pollutants. The understanding of its chemical behavior in environments is important to prevent the recurrence of the Minamata tragedy.

Beside industrial disposal and anthropological origin through the burning of coal, mercury is also introduced into environments through volcanic and hydrothermal activities. These multiple emission sources of mercury make it daunting to resolve its pathways in geochemical cycles in nature. To make matters worse, each pathway involves various types of mass-independent isotopic fractionation besides a common mass-dependent fractionation. As in the case of sulfur, however, development in the study of mass-independent isotopic fractionation processes now sheds light on this problem.

Thanks to the large number of stable isotopes of mercury (seven isotopes), we are now able to disentangle extremely complicated pathways of mercury in nature, providing a very powerful means to control this highly toxic element in nature.[5] Although fundamental processes in individual mass-independent isotopic fractionation are not fully understood yet, multiple mass-independent fractionation processes are now being used to trace the geochemical cycles of several elements including cadmium, which is another toxic element.

STARDUST

In concluding this chapter, we introduce an astounding example of the use of an isotope as cosmic DNA. As already mentioned, the isotopic compositions of elements in the Earth reflect only the average solar nebula composition, but we can actually find out in what kind of a star some specific isotopes were born by studying in great detail the isotopic composition of meteorites.

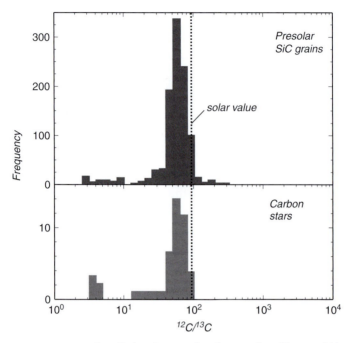

FIGURE 8.2 Carbon isotope data for presolar silicon carbide grains (top) and carbon stars (bottom).[8]

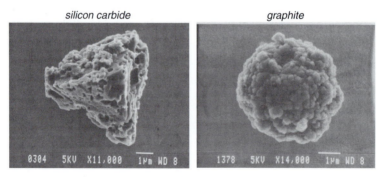

FIGURE 8.3 Photographs of stardust gains (courtesy of S. Amari).

In Figure 8.2, we show carbon isotopic ratios observed by spectroscopy in some carbon stars. Carbon stars are a peculiar type of red giant stars, characterized by a high abundance of carbon (more abundant than oxygen). The same figure also shows the carbon isotopic ratios of tiny inclusions separated from a primitive meteorite, measured by Sachiko Amari at Washington University. The two carbon isotope ratios agree almost perfectly with each other, giving a robust proof that the tiny inclusions did indeed come from carbon stars. Such tiny inclusions of presolar origin are called stardust. Since the first separation from a primitive meteorite by the Chicago group almost two decades ago, a number of stardust grains have been reported in a variety of meteorites, whose host stars are identified as a variety of stars in addition to carbon stars.[6]

Extracting extremely tiny pieces of stardust, commonly of a few micrometers in size or smaller, from a bulk meteorite was a difficult task. Edward Anders, who initiated the separation project at the University of Chicago two decades ago, likened this formidable work to finding a needle in a haystack. This reminds us of the legendary work of Marie and Pierre Curie, who succeeded in extracting a tiny amount of uranium from pitchblende ore loaded in a freight car.

Figure 8.3 shows electro-microphotographs (images taken through an electron microscope) of typical stardust grains, which were nicknamed cauliflower and onion by Amari. We can now literally see under a microscope and even touch tiny pieces that have traveled over many trillions of kilometers in the universe from various stars after their birth more than 4.5 billion years ago (the age of the host meteorites).

NOTES

1 For introductory materials on climate science, see [9, 10].

2. Note that in the original report [11] the observed deviation from ordinary mass-dependent fractionation was interpreted to reflect some nucleosynthetic processes.

9   The Earth's magnetism

MAGNETIC MINERALS IN ROCKS: SMALLER IS BETTER

As mentioned in Chapter 5, paleomagnetic studies suggest that remanent magnetization imprinted on volcanic rocks has been preserved for hundreds of millions of years without changing its direction and intensity. How can such remarkable stability be possible? To answer this question, we first need to introduce some basic concepts of rock magnetism. Magnetic materials such as iron, nickel, and iron oxide (typically magnetite) are said to be ferromagnetic, meaning that when subjected to an external magnetic field, they are magnetized strongly in parallel to the magnetic field. Even after the removal of the applied magnetic field they still retain a noticeable amount of magnetization, which is called remanent magnetization.

Ferromagnetism is retained only below the critical temperature known as the Curie temperature. This phenomenon was discovered by the French physicist Pierre Curie (the husband of Marie Curie). For example, the Curie temperatures of iron, nickel, and magnetite are 770 °C, 358 °C, and 585 °C, respectively. Above the Curie temperature, ferromagnetism disappears, and these materials no longer exhibit magnetization.

If a ferromagnetic material such as an iron particle is large enough (e.g. a few tens of micrometers at room temperature), it is divided into magnetic domains, partitioned by magnetic domain walls a few tens of nanometers in thickness (Figure 9.1). Each domain has permanent magnetization of its own, but its orientation is random so that the net magnetization of a bulk body is zero. This can be understood on the basis of the minimum energy principle: a minimum magnetic energy state of a ferromagnetic body is achieved when the net magnetization is null. When the bulk body becomes smaller, however,

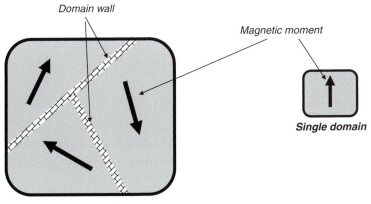

Domain wall

Magnetic moment

Single domain

Multiple domains

FIGURE 9.1 A ferromagnetic particle (e.g. iron at room temperature) larger than a few micrometers is divided into magnetic domains, resulting in the reduction of magnetic energy.

the formation energy of magnetic domain walls surpasses the energy reduction due to the null magnetic state, and the bulk body is no longer divided into magnetic domains and becomes a single-domain particle. The critical size for a single-domain magnetite grain is a few microns or less at room temperature. Magnetite grains in volcanic rocks are generally close to or less than the critical size, and generally regarded to behave as single-domain particles.

In the case of volcanic rocks, magnetite is a major ferromagnetic mineral. Magnetite is dispersed in a volcanic rock matrix as tiny grains, generally smaller than one micron, because of rapid cooling at the Earth's surface. Therefore, most of the magnetite grains in volcanic rocks are small enough to become single-domain particles and carry permanent magnetization. If magnetite grains are larger than a few tens of microns, they will be divided into magnetic domains and will not carry remanent magnetization. In contrast to volcanic rocks, plutonic rocks such as gabbro and granite that are crystallized very slowly in the deeper crust contain much bigger multi-domain magnetite crystals, and therefore they hardly acquire any remanent magnetization. This leaves only quickly chilled volcanic rocks usable for paleomagnetic studies, although both plutonic and volcanic rocks generally contain a similar amount of magnetite.[1]

THERMO-REMANENT MAGNETIZATION CAN PERSIST FOREVER

When discussing the evolution of the Earth, we are dealing with a time scale up to 4.5 billion years. Can remanent magnetization be stable over such an enormous time span? The answer is yes. Apart from the physical preservation of the host volcanic rock, remanent magnetization once acquired by volcanic rocks during the cooling of molten lava to room temperature (known as thermo-remanent magnetism) is shown to be stable for more than the age of the Earth, as long as it has been kept well below the Curie temperature. This outstanding characteristic was worked out theoretically by Louis Néel in France, a 1970 Nobel laureate in physics.[2] Néel pointed out that the main cause of diminishing thermo-remanent magnetization was thermal agitation. He showed that the time scale for thermo-remanent magnetization to diminish (called the relaxation time) is proportional to an exponential function of temperature, $\exp(a/T)$, where T is temperature and a is a constant depending on material properties. The exponential function increases extremely rapidly even with a small increase in its argument. For example, $\exp(7)$ is about one thousand, but $\exp(14)$ is about one million. Whatever the value of the constant a is, therefore, the term $\exp(a/T)$ quickly diverges to infinity as temperature drops, and so does the relaxation time.

Freshly erupted lava is well above 800 °C, which is much higher than the Curie temperature of magnetite. When it cools down past the Curie temperature, magnetite becomes ferromagnetic and acquires magnetization. The individual magnetite grains align their magnetization in the direction of the ambient magnetic field at the time. The intensity of this acquired magnetization is proportional to the intensity of the ambient magnetic field. When the lava approaches room temperature, the relaxation time becomes exponentially long, easily exceeding the age of the Earth. That is, once the remanent magnetization is acquired, it is no longer susceptible to the ambient magnetic field at room temperature. This means that a memory of the ancient geomagnetic field at the time of lava's eruption, in both direction and intensity, is frozen in the volcanic rock permanently as a thermo-remanent magnetization.

GENERATION OF THE GEOMAGNETIC FIELD BY GEODYNAMO

The geomagnetic field, if averaged over millions of years, appears like a simple "dipole" field produced by a huge bar magnet, having poles at the Earth's north and south poles. If the geomagnetic field is studied in more detail, however, it changes relentlessly in its direction and intensity with time scales ranging from sub-second to hundreds of millions of years. We briefly mentioned in Chapter 3 that the geomagnetic field is generated within the liquid metallic outer core, and its ceaseless variations are attributable to the complex fluid motion in the outer core. The mechanism is analogous to a huge electromagnetic generator (a dynamo) in a power station. The mechanical motion of an electrically conductive fluid, in a seed magnetic field originating in the Sun, results in electromagnetic induction to generate electric currents in the fluid, which in turn gives rise to an intense planetary-scale magnetic field.

To describe the process more precisely, we need to solve two basic equations simultaneously. One is the Navier–Stokes equation to describe the motion of fluid in the outer core, and the other is the Maxwell equation to relate electromagnetic processes. Both equations are among the most profound and fundamental equations in modern physics. Solving even one of them is challenging enough, but to understand the origin of the geomagnetism we must solve both equations simultaneously! The discipline is named magnetohydrodynamics, which was first systematically studied at the University of Cambridge under the direction of Edward Bullard. One of the pioneers in this new discipline was Subrahmanyan Chandrasekhar, a 1983 Nobel laureate in physics.

One of the authors, Minoru Ozima, did a postdoctoral study in Chandrasekar's group and had the interesting experience of witnessing Chandrasekar's attitude to research.[1] One Thursday afternoon in the early 1960s, Chandrasekhar made his regular once-a-week visit to a seminar room in the downtown campus of the University of Chicago. He had an office at the Yerkes Observatory about 100 miles north of Chicago. All students were attentive, in the expectation of hard questions from Chandrasekhar. Ozima was also in the seminar room, and a student sitting next to him happened to be fussing about the difficulty of his assignments on magnetohydrodynamics. Chandrasekhar's advice was the following: "Just look at the equations, nothing but look, look, look, for at least three days. Then you will find that they start to talk to you." Chandrasekhar himself made a number of significant contributions to a variety of problems such as the structure and evolution of stars, by providing elegant analytical solutions to mathematically challenging equations.[2] In the field of magnetohydrodynamics, however, this "look hard and get the equations to talk back" approach, a traditional method to solve equations analytically, has largely been replaced by a computational approach, namely conducting numerical simulation with the help of supercomputers. This is a reasonable transition, because magnetohydrodynamics is an extremely nonlinear and complex process, the detailed understanding of which is well beyond the capacity of analytical methods. Even the fastest supercomputer available today, however, still falls short of realistically simulating the processes operating within the Earth's core. Researchers have been able to numerically simulate a self-sustaining geodynamo, and even the occasional reversal of its magnetic polarity, but the parameter range that can be handled by current computer technology is still too limited to approach an Earth-like regime.[3]

MANTLE CONTROL OVER THE HISTORY OF THE GEOMAGNETIC FIELD

The mathematical difficulty described above is not the only reason that the theoretical study of the geomagnetic field generation in the core is challenging. The convective fluid motion in the core, which is essential for a geodynamo, is driven fundamentally by mantle cooling. At the very least, the mantle must be colder than the core. If, instead, the mantle happens to be hotter than the core, there is no reason for convection to take place within the liquid outer core, and therefore no geomagnetic field can exist. Because we do observe a geomagnetic field today, we are certain that the mantle is sufficiently colder than the core to drive the geodynamo, but has the mantle always been colder than the core? If so, how much colder? As you may realize, if we want to know what was happening in the core in the past, we also need to understand the state of the mantle at the same time. That is, if we want to understand the core, we need to look at a wider scale and think how the entire solid Earth was cooling over the geological time scale.

As explained in Chapter 1, the cooling rate of a planet depends on the balance between internal heat production and surface heat loss, and for the present-day Earth, internal heat production is about half of surface heat loss. Internal heat production, however, cannot be constant over time. Was it higher or lower in the past? It has always been higher in the past because internal heating is produced by radioactive decay. The number of radioactive nuclei is monotonically decreasing with time, so the internal heat production is too. As we go backward in the past, therefore, internal heat production must have been higher, and given the half-lives of major heat-producing isotopes (^{238}U, ^{235}U, ^{232}Th, and ^{40}K), we can estimate that it was about twice as high at about 3 billion years ago. What can we say about the rate of surface heat loss at that time? This turns out to be difficult to answer. The rate of surface cooling depends on the vigor of mantle convection, and there is still much debate about how vigorously the mantle was convecting in the past.[4] A naive expectation is that the mantle should have been convecting more vigorously in the past because the Earth was hotter, but geological observations suggest otherwise, and surface heat loss may have been more or less constant. If surface heat loss at 3 billion years ago occurred at a similar rate to the present-day value, for example, internal heating and surface heat loss would have been perfectly balanced, which in turn means that the mantle was not cooling around that time. This does not necessarily mean that the core was also not cooling, because it is still possible for the mantle to have been colder than the core, but this discussion on planetary cooling illustrates a theoretical difficulty arising from the interconnected nature of the evolution of the Earth system. By the same token, however, this interconnectedness is what makes certain kinds of observations extremely valuable. If, for example, we can figure out when the geodynamo started to operate in Earth's history, it will help us better understand not only the evolution of the core but also that of the mantle. This is the topic we will examine in some detail next.

ONSET OF THE GEOMAGNETIC FIELD IN THE EARTH'S HISTORY

When did the geomagnetic field first appear on the Earth? The oldest volcanic rocks that have been studied so far for paleomagnetism are komatiites from the Komati river region in South Africa, estimated to be about 3.5 billion years old.[5] These volcanic rocks appear to show stable remanent magnetization. However, it is difficult to conclude whether or not the volcanic rocks have faithfully retained the original remanent magnetization acquired 3.5 billion years ago. One specific concern is that they might have undergone metamorphic alteration so that the remanent magnetization had been reset at some later period, either thermally or chemically.

A paleomagnetic study is commonly made on a hand-sized specimen of volcanic rock. Although a volcanic origin (i.e. being cooled very rapidly) assures that the acquisition time of remanent magnetization is nearly coincident with the eruption time of volcanic rocks, the thermoremanent magnetization can still be affected by the geomagnetic field over a long geological time. As mentioned above, this problem may be safely discarded if the magnetic mineral carrying the remanent magnetization is of a single-domain size, but magnetite grains in some volcanic rocks buried beneath a thick volcanic lava flow may have become larger than a single-domain size owing to slower cooling. How can we choose only single-domain grains for a paleomagnetic investigation?

Recently, John Tarduno of the University of Rochester has achieved a breakthrough to this challenging problem.[6] With the use of a superconducting quantum interference device, he succeeded in measuring the thermo-remanent magnetization imprinted on a minute magnetite grain, a few tens to a few hundred nanometers in size, that had been encapsulated in a silicate mineral grain such as quartz or feldspar, a few hundred microns in size. The key to this outstanding accomplishment is that the carrier of remanent magnetization, the magnetite, is of an extremely small size (less than a few hundred nanometers), which is well below the critical size of a single domain of magnetite and hence assures its high magnetic stability, even exceeding the age of the Earth, as we have already learned. The additional beauty of this elegant technique is that owing to encapsulation in a silicate mineral, the magnetite was fairly well protected from metamorphic alteration and so retained its original thermo-remanent magnetization. Taking full advantage of these special conditions, the acquisition time of remanent magnetization can be equated with the crystallization age of a host plagioclase grain, which can be measured by means of isotopic dating. Tarduno and his colleagues concluded that robust thermo-remanent magnetization was retained in single-domain magnetite grains encapsulated in silicate crystals of 3.2-billion-year-old rocks from South Africa. If this remanent magnetization were assumed to have been acquired under the influence of a geomagnetic dipole field, we further infer that its dipole intensity was similar to the present geomagnetic dipole. This observation may suggest that the intensity of the geomagnetic field has been stable over the past 3.2 billion years.

However, the existence of robust remanent magnetization measured at a few sampling sites does not necessarily prove that there was a global geomagnetic field similar to the present geocentric axial dipole field. For example, it is well known that lightning can give rise to a strong local magnetic field. Meteorite impacts can induce regional magnetic anomalies of a much larger scale, extending over a few hundred kilometers, some of which are now well identified. In order to conclude the emergence of the global geomagnetic field, therefore, systematic paleomagnetic studies covering vast extents of time and space are required. In the next chapter, we introduce a totally new approach on the basis of observation of soils on the lunar surface, which has enabled us to infer ancient records not only of the geomagnetic field, but also some of the most fundamental issues in planetary sciences, such as the time of the first appearance of oxygen in the atmosphere.

NOTES

1. Ozima started as a geophysics student working on rock magnetism, in particular mineral dating with the potassium–argon method, and received a Ph.D. in 1958 from the University of Toronto under the supervision of R. D. Russell. He then spent about a year at the University of Chicago for a postdoctoral study in Chandrasekhar's group and worked on the hydromagnetic stability of Taylor–Couette flow.[7, 8] He remained active in the field of rock magnetism for a while, but in the early 1970s he turned his attention to the geochemical aspect of the potassium–argon system and has since expanded his expertise into noble gas geochemistry.

2. Chandrasekhar chose to work continuously in one specific area of physics, typically for more than several years, and published his findings in a compendium at the end.[9–11] His work on hydrodynamic and hydromagnetic stability[11] is one of the classics of fluid mechanics.

10 The Moon: a looking glass to mirror the ancient Earth

GEOCHEMICAL AND GEOPHYSICAL FOSSILS

In tracing the evolution of the Earth, instead of using physical and chemical methods to reproduce phenomena in the laboratory, research is carried out using the "fossils" of these phenomena. We have already explained how the isotopic ratios recorded in rocks, the oceans, and the atmosphere serve effectively as the "fossils" of ancient chemical differentiation events. Meteorites are still the most important source of information on the origin of the Earth, being "fossils" in which the early state of the Solar System is frozen. We have also seen that the remanent magnetism of rocks can be used as a "fossil" of the Earth's magnetic field in the past. Studying the evolution of the Earth, therefore, often calls for an ingenious approach, requiring more than simple applications of mathematical, physical, and chemical methods. In this chapter we will discuss a novel approach in this spirit, which may be called "lunar paleopedology" (pedology is the scientific discipline of studying soil). After the monumental achievement of the landing on the Moon in 1969 under the Apollo project, lunar soils sampled by astronauts have been offering a new type of "fossil record" not only for the Moon, but also for the Sun and Earth. We start with a short discussion of the origin of the Earth–Moon system.

ORIGIN OF THE EARTH–MOON SYSTEM

Although the origin of the Moon is still a big mystery, a currently predominant view is that the Moon formed as a result of the impact of a Mars-sized proto-planet with a proto-Earth.[1] An almost identical isotopic composition of oxygen (the most abundant element in planetary objects) between the Moon and Earth suggests that both objects had undergone elemental and isotopic homogenization, most likely through high-temperature vaporization caused by the impact. Numerical calculations of such a giant impact have commonly indicated that the temperature in the impacted material (a proto-Earth) and impactor could have reached more than ten thousand degrees centigrade, enough to vaporize all of the materials involved.[2] The formation age of the Earth–Moon system (or the timing of the giant impact) has been estimated to be about 4.5 billion years ago from an extinct short-lived radioactive isotope ^{182}Hf, whose half-life is 9 million years (see Chapter 2).

Theoretical studies on the dynamics of the Earth–Moon system, under the Sun's gravitational force, suggest that a newly born moon was much closer to the Earth, and that the Moon has gradually been receding to its present position, which is a distance equivalent to about 60 Earth radii. Because of the viscous nature of the ocean, a high tide (or the maximum vertical uplift of ocean water) always occurs after the Moon has passed the bulge of the ocean, resulting in a torque force to accelerate the revolution of the Moon (Figure 10.1).[3] As the Moon revolves faster, it recedes further from the Earth.

The receding of the Moon from Earth has indeed been observed by measuring the travel time of laser light between the Earth and Moon with mirrors left on the Moon by Apollo astronauts. The receding rate deduced from that measurement over the past few decades is 3–4 cm/yr. The rate in ancient times must have been higher than this current rate, since the likely shorter distance between the Earth and the Moon means that tidal interaction must have been much greater in the past. Figure 10.2 shows a theoretical calculation of the variation of the Earth–Moon distance over time. In this calculation, the Earth was assumed to have the same amount of ocean water (the principal torque-generating agent) as at present throughout its history. This assumption may be justified at least back to about 4 billion years ago, judging from the wide occurrence of sedimentary rocks in the crust.

Dynamic calculations also suggest that because of the huge mass difference between the Earth and Moon, the synchronous rotation, i.e. the spinning of the Moon being nearly identical to its revolution period

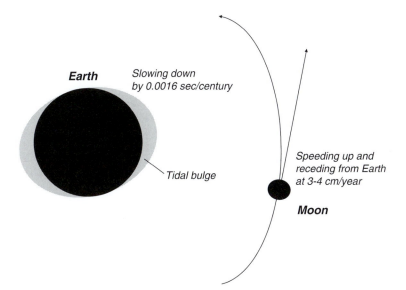

FIGURE 10.1 The highest tide would occur on the side of the Earth facing the Moon, but because of viscous coupling between the oceans and the solid Earth, it moves away from the Moon quickly as the Earth rotates. Because of the gravitational interaction of the tidal bulges between the Earth and Moon, this time lag exerts a force to speed up the revolution of the Moon around the Earth, on the one hand, and to slow down the Earth's rotation, on the other hand.[3]

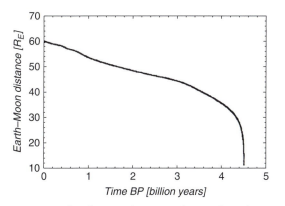

FIGURE 10.2 The distance between the Earth and Moon has been increasing with time. About 4 billion years ago, it is estimated to have been about half the present distance.[12]

around the Earth, was established very quickly (possibly on a time scale of years) after the formation of the Earth–Moon system.[4] Owing to this synchronous rotation, theoretical calculations further predict that the far side of the Moon has never faced towards the Earth since its birth about 4.5 billion years ago. That is, any living organisms throughout the Earth's history must have seen only the near side of the lunar surface, as we are seeing today.

Although the above theoretical predictions on the dynamics of the Earth–Moon system do not generally contradict our current geological information, neither has their accuracy been verified. Indeed, making a theoretical prediction and validating (or invalidating) it with observations is how science progresses in general. Testing predictions is, however, very difficult for the time of the early Earth, especially for the period before 4 billion years ago, because no rock has been preserved from this time. Recently, a totally new way to study the Earth–Moon system has emerged, which may also hold a key to testing these theoretical predictions. This approach is to use lunar soils as "fossil records" for the evolution of the Earth–Moon system. We start from the introduction of the solar wind, which is the key messenger of fossil records in this new approach.

SOLAR WIND ON LUNAR SOILS

The Sun has been continually ejecting matter into space, and this is known as the solar wind. The solar wind consists mainly of hydrogen and helium ions, with a trace amount of heavier ions, and travels with a mean speed of a few hundred kilometers per second. Although the existence of the solar wind is well confirmed by spacecraft observations, its origin is still not fully understood.[5] Currently, it is assumed that high temperatures at the bottom of the solar atmosphere push the solar wind outward. The existence of the solar wind has been traced back almost to the beginning of the Solar System from analyses of lunar soils of various surface exposure ages. Based on lunar soils, a solar wind of roughly similar intensity and of similar elemental composition can be traced back to more than 4.2 billion years ago.

When the solar wind encounters the Moon, it directly impinges onto the lunar surface, because there is practically no atmosphere to prevent its intrusion. The solar wind particles are then implanted onto surface mineral grains that happen to be at the very top of the soil. These surface exposed grains are about 10 micrometers in size, and numerical modeling suggests that these grains have a few hundred thousand years of mean residence time on the very top surface, after which they are gradually covered by dust grains continuously supplied by subsequent bolide impacts (as evidenced by numerous lunar craters).[6] Therefore, after being buried, lunar soils keep the implanted signature of the solar wind of their surface exposure time, which we can use as "fossils" of the solar wind. From this fossil record, we can resolve the evolutionary history of the Earth–Moon system back to 4.2 billion years ago.

THE EARTH WIND: IONS ESCAPING FROM THE EARTH

Unlike the Moon, the Earth's surface is well shielded from the solar wind not only by a thick atmosphere, but also by the geomagnetic field. When the solar wind particles (fully ionized charged particles called plasma) approach the Earth, they will be deflected away from it by the geomagnetic field. This is schematically illustrated in Figure 10.3a. The approaching solar wind is prevented from direct intrusion into the Earth's atmosphere, and passes around the Earth. From the balance between the kinetic force of the impinging solar wind attempting to penetrate into the Earth's atmosphere and the geomagnetic repulsive force that resists its intrusion, we can estimate the distance at which the solar wind starts to be deflected from the Earth.[7] This distance is about ten Earth radii towards the Sun.

Although the intrusion of the solar wind into the atmosphere is prohibited, a small fraction of the atmospheric constituents, essentially oxygen (O^+) ions because they ionize relatively easily, can be picked up from the upper atmosphere by the stream of the passing solar wind. These Earth-escaping oxygen ions were indeed confirmed by a spacecraft mission named GEOTAIL,[8] and we may call the escaping atmospheric

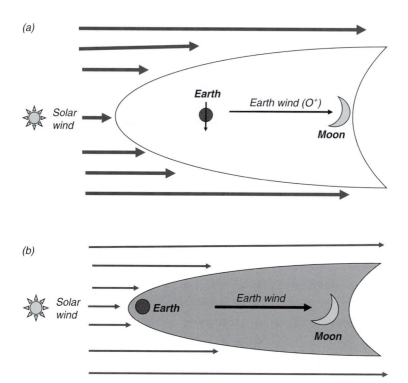

FIGURE 10.3 Solar wind. (*a*) Solar wind on a magnetic Earth. The Earth is shielded from the solar wind by the geomagnetic field, enveloped by a magnetosphere. A measurable amount of oxygen ions (Earth wind, O^+) are picked up by the solar wind, of which about 10% reaches the lunar surface. (*b*) Solar wind on a non-magnetic Earth. If the Earth has no global geomagnetic field, the solar wind approaches the Earth more closely, and a stronger Earth wind (essentially oxygen ions if oxygen was already present in the atmosphere) hits the Moon.

components the Earth wind as opposed to the solar wind. The amount of oxygen ions in the Earth wind measured at the lunar orbit by the GEOTAIL mission is similar to that of oxygen delivered by the solar wind onto the Moon. Recently, Trevor Ireland of Australia, Ko Hashizume of Japan, and Marc Chaussidon of France have successfully measured oxygen implanted into lunar soils.[9, 10] They found a notable isotopic deviation from the solar wind oxygen. This non-solar oxygen may be attributable to the Earth wind as indicated by the GEOTAIL mission. If the non-solar oxygen is confirmed to be of terrestrial atmospheric origin, the first appearance of such oxygen in lunar soils must coincide with the appearance of biogenic oxygen in the atmosphere.

GEOMAGNETIC FIELD MIRRORED ON THE MOON

The interaction of the solar wind with the Earth is primarily controlled by the geomagnetic field. This leads to another intriguing inference. As we discussed in Chapter 9, we are still far from a complete understanding of the origin of the geomagnetic field, especially regarding the time when the geomagnetic field first appeared on Earth. From a systematic examination of lunar soils for ancient solar wind records, however, we can directly tackle this problem.

To explain this, let us suppose that once upon a time, there was no magnetic field on the Earth. The solar wind would then approach the Earth more closely without the resistance of the geomagnetic field until it was finally stopped by the atmospheric pressure. Again from considering the balance between the solar wind pressure and the atmospheric pressure, we infer that the solar wind will be stopped at about a few hundred kilometers above the Earth's surface on the Sun-facing side (Figure 10.3b). Because the solar wind can approach the Earth more closely, more atmospheric constituent materials will be picked up by the solar wind than when the geomagnetic field is present (Figure 10.3a).

Therefore, the amount of solar wind that is implanted onto lunar soils will vary depending on whether the Earth has a planetary-scale magnetic field. If we find a systematic enrichment in the amount of solar wind implanted into lunar soils, for example, we may relate this period to the absence of the geomagnetic field. From further systematic and Moon-wide studies into the implanted solar wind, we may resolve the onset time of the Earth's magnetic field.

Recent precise isotopic analyses of several trace elements such as nitrogen, hydrogen, and noble gases appear to be promising in this regard. Hashizume and his colleagues observed large variations in the isotopic compositions of hydrogen and nitrogen implanted by the solar

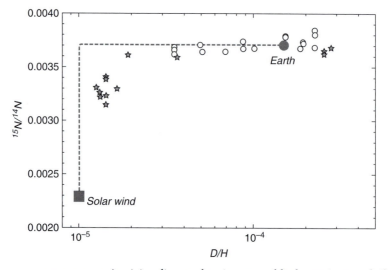

FIGURE 10.4 A mixing diagram for nitrogen and hydrogen isotopes.[13] Ancient lunar grains (stars) show in general a higher fraction of the solar wind component, whereas younger grains (circles) are more abundant in the Earth component.

wind in lunar soils, and found that the isotopic variation in hydrogen and nitrogen exceeded a few tens of percent, which was difficult to attribute to ordinary isotopic fractionation processes.[11] What is even more interesting is the fact that the deviation from the solar wind isotopic component is in general larger in ancient soils (older than 3.8–3.9 billion years ago) than in younger soils (a few hundred million years old) (Figure 10.4).

Moreover, when the data were plotted in a two-component mixing diagram (Figure 10.4), the scatter in the isotopic compositions was consistent with a mixing curve constructed for the solar wind component and the Earth's atmospheric component. All of these observations can be consistently explained if there was no geomagnetic field during the period when the older lunar soils were exposed to the solar wind about 3.8–3.9 billion years ago.

Since the Moon is supposed to have acquired its spin-lock status soon after the formation of the Earth–Moon system, we do not expect any Earth wind signature to be implanted on the Moon's far side at all. Thus, if future lunar missions acquire soils from the far side of the Moon, lunar paleopedology may also resolve the question of when the Earth–Moon system acquired the synchronous rotation, for which we have only theoretical predictions so far. Thus, we see here another unique example of "geophysical fossils" in lunar soils. The lunar soils may be likened to a looking glass to mirror the Earth over more than 4 billion years.

11 The past and future of the evolving Earth

USEFULNESS OF THE EARTH'S HISTORY
The Earth's history is in itself a very interesting research theme, but it is also essential if we want to understand fully what is happening in the present Earth. For example, let us consider the Earth's magnetism. As said in earlier chapters, the fluid motion within the metallic core acts as a generator and gives rise to the geomagnetic field. Studying magnetohydrodynamics in the present-day core alone, however, will not lead to a total understanding of the origin of the geomagnetic field, because many of its characteristics, such as the reversals of its magnetic polarity, need to be examined on a time scale of tens of millions of years. In order to understand why and how the geomagnetic field emerged, we need to understand not only the evolution of the core, but also the evolution of other components in the Earth system. In fact, most topics in Earth sciences can be understood more deeply in the context of the evolving Earth. The Earth is steadily cooling down with declining internal heat production, and everything is changing with time, albeit very slowly.

Studying the Earth's history also provides the most effective means to forecast the fate of this planet. We cannot, of course, predict everything by studying the past. For example, research into the Earth's history does not give us a particularly useful way to predict the timing of major earthquakes, which may occur anytime. Also, studying the past does not instantly provide a clue to what we should do for the future. History does not repeat itself in exactly the same way. A careful reconstruction of what happened in the past must be followed by a theoretical study to understand why and how exactly it happened. Without this combination of observations and theory, we cannot extrapolate our understanding to the future Earth with confidence.

Since the late twentieth century, Earth sciences have become an essential scientific discipline regarding the fate of human civilizations. As Thomas Robert Malthus predicted near the end of the eighteenth century,[1] the human population has grown approximately exponentially, and the world population is now about 7 billion. There are no more frontiers to migrate to, and human beings have begun to feel the finite nature of the Earth's surface. We have also realized that human activities can have global impacts, often in an unexpected way because of the complex behavior of the Earth system. Studying the Earth's history over a time scale of tens of millions of years is not merely a subject of intellectual interest, but is of fundamental importance in considering the immediate future of the habitable planet Earth. Nuclear waste disposal and greenhouse agents in the atmosphere are among the most urgent examples.

THE EARTH'S EVOLUTION, PAST AND FUTURE

To hand a habitable planet over to the next generations, we need to consider its fate not only on time scales of a few tens of years, comparable to our own life spans, but also on a time span of thousands of years or more, so that our descendants will be well prepared for such formidable problems as global warming and the disposal of radioactive waste. The study of the Earth's past climate is called paleoclimatology, and as briefly mentioned in Chapter 8, the variation of paleotemperature over the geological time scale as inferred from oxygen isotopic data provides extremely valuable constraints on how the complex climate system works. Even though anthropogenic carbon emissions by the burning of fossil fuels are a recent and also unique event in the Earth's history, the concentration of carbon dioxide in the Earth's atmosphere has varied considerably in the past, and it was much higher than the present-day value during most of the Phanerozoic (i.e. the past 540 million years).[2] Resolving how the Earth's environments responded to such a high concentration of carbon dioxide is one of the important themes in paleoclimatology. The burning of fossil fuels will eventually cease, probably in the next hundred years or so, because of the exhaustion of fossil fuels themselves, and the increased amount of carbon dioxide in the atmosphere will mostly be absorbed by the oceans with a time scale of a few hundred years.[3] As such, the time scales involved are relatively short in a geological context, though the outcome of such a rapid rise in atmospheric carbon dioxide could have disastrous consequences for a rapidly rising human population and for the diversity of life on our planet.

The problem of how to dispose of radioactive waste is no less serious, and is challenging in a very different way because the time scales involved are potentially much longer. Here, let us see a unique approach based on geochemical "fossils" for radioactive waste disposal.

RADIOACTIVE WASTE DISPOSAL

Since the oil crisis in 1973, the issue of future supplies of energy has been discussed practically daily in the mass media. It is evident that oil and other natural resources are limited, and everyone recognizes that we need to seek alternative energy resources. The burning of fossil fuels should also be reduced because of its impact on the global climate. As one of the most promising forms of energy, nuclear power has been frequently discussed because it does not release carbon dioxide (although the uranium resource will not last for another century if we choose to increase nuclear power generation).[4] In addition to the problem of safety, however, the use of nuclear power also involves the extremely difficult problem of disposal of the radioactive waste continually produced by nuclear reactors. In order to solve energy problems in the near future, we must deal with this issue first. In fact, even if we do not choose to use nuclear power in the future, we have been using it for the past 50 years, yielding a tremendous amount of radioactive waste in many countries, none of which has found a reliable solution for its disposal.[5]

Among various methods so far proposed, burying radioactive waste underground, in a geological setting for which we can be assured of tectonic stability and resilience against geochemical erosion, seems to be the most practical way. A difficult part is how we can be sure of such properties over a sufficiently long period. The Oklo "natural reactor" provides an excellent test case for this problem. The discovery of the Oklo natural reactor is not only a fascinating story; it also demonstrates the importance of a geochemical "fossil" approach.

OKLO NATURAL NUCLEAR REACTOR

In 1972, French scientists reported that uranium from the Oklo mine in the Republic of Gabon (Figure 11.1), close to the Atlantic coast of Africa, had an extremely unusual isotopic composition, and they suggested that the cause of this unusual isotopic ratio was a nuclear chain reaction involving uranium that occurred at this site about 1800 million years ago.[6] When a chain reaction takes place, the isotopic composition of uranium changes. It is as if a nuclear reactor power station were operational at that site. Figure 11.2 shows the measured values of the isotopic ratio of uranium ores from the Oklo mine. The isotopic ratio of uranium-238 and uranium-235 varies widely by as much as 40 percent, a remarkable contrast to the almost uniform values of isotopic ratio obtained from elsewhere. The only reasonable explanation for such a large abnormality in the isotopic ratio is a nuclear reaction. Actually, the possibility of a natural nuclear chain reaction had already been seriously considered soon after the success of a prototype nuclear reactor underground at the University of Chicago campus in the early 1940s.

In order for a nuclear chain reaction to occur, uranium-235 must be sufficiently enriched. As explained in Chapter 1, naturally occurring uranium is composed of two isotopes, uranium-235 and uranium-238, and only uranium-235 is fissile, i.e. capable of sustaining a chain reaction of nuclear fission. When a neutron is absorbed by the nucleus of a uranium-235 atom, the atom will split into fast-moving lighter elements and free neutrons, and if one of these free neutrons is absorbed by another uranium-235 atom, this fission reaction will continue. Uranium-235, however, makes up only about 0.7 percent of naturally occurring uranium, and no matter how much of this natural uranium is collected, a chain reaction will not occur. In order to trigger a nuclear chain reaction, the concentration of uranium-235 must

FIGURE 11.1 An aerial photo of an Oklo uranium mining pit (a few hundred meters in scale) in the Gabon Republic, Central Africa. In the middle of the side cliff above a small lake, tens of meters across, at the bottom of a pit, several severely altered lithological spots (a few square meters) with radiogenic isotopes characteristic of uranium nuclear chain reactions can still be recognized. (Courtesy of H. Hidaka and F. Gautier-Lafaye.)

exceed a few percent. Uranium-235 has a half-life of approximately 700 million years, whereas uranium-238 has a longer half-life of about 4.5 billion years. If we go back in time, the amount of uranium-235, which has a shorter half-life, will increase more rapidly compared with

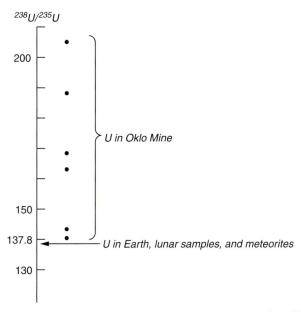

FIGURE 11.2 Isotopic ratio of uranium (U) in the Oklo mine, Republic of Gabon. The isotopic ratio (^{238}U/^{235}U) varies greatly in comparison to the almost identical value of 137.8 in terrestrial and lunar samples as well as meteorites.

that of uranium-238. A billion years ago the concentration of uranium-235 must have been about 2 percent, which is close to uranium fuels being used for a commercial nuclear reactor.

In addition to the uranium-235 enrichment, another essential condition for the occurrence of a chain reaction is the existence of water (or more precisely, hydrogen in the form of water) around the uranium. This is because when the fast-moving neutrons released from uranium collide with hydrogen atoms in the water, they slow down, becoming more easily absorbed by another uranium atom and thus facilitating a chain reaction. Taking these facts into consideration, in the early 1950s scientists concluded that, even though uranium-235 was probably sufficiently enriched more than 2 billion years ago, there would not have been enough hydrogen atoms around a uranium ore to reduce the speed of the neutrons, preventing a chain reaction in nature.[7]

Around that time, however, Paul K. Kuroda of the University of Arkansas made a careful calculation on the geological feasibility for nuclear chain reaction in nature in the remote past. He came to the conclusion that if there was sufficient ground water as a source for hydrogen around uranium ore minerals, and if the uranium deposit had a suitable scale of a few tens of centimeters or larger, a uranium deposit could have reached a critical condition for a chain reaction about 2 billion years ago. It was amazing that the geological setting of the Oklo mine was so close to the one predicted by Kuroda. Kuroda's work met with considerable skepticism, and it took a while for his seminal study to be published in an academic journal.[8] Sixteen years later, however, his remarkable prediction was fully confirmed by the discovery of the Oklo natural reactor. To date, no other "natural reactors" besides the Oklo mine have been discovered.

OKLO NATURAL REACTOR AS A "FOSSIL"

The fact that no other natural reactors have been found indicates either that the geological conditions in the Oklo mine were unique in their ability to trigger a chain reaction or that the geological setting in the environment was especially resistant against later geological disturbances such as weathering or hydrothermal leaching. Since there is no definite evidence to rule out either of them, both factors are likely to have contributed to the existence of the unique natural reactor. Whichever is the case, this ancient natural reactor provides us with an excellent "fossil", from which we can assess the fate of highly hazardous radioactive waste over a geological time scale.

Like a commercial nuclear reactor, the Oklo reactor has produced a large amount of radioactive products. By studying the isotopic compositions of these waste products in the reactor zone and its surroundings, scientists have estimated that a nuclear chain reaction quite similar to that generated by a current commercial nuclear reactor was intermittently operating for a few hundred thousand years, and that the energy generation rate was about 100 kilowatts on average throughout this period.[9] The energy generation rate is four orders of magnitude lower than that of a medium-size commercial reactor, but because the duration is so long, the total energy released in the reactor zone would have been similar to that produced by a medium-size reactor over a few tens of years. The amount of nuclear waste produced by the Oklo reactor is thus comparably large.

With the use of these "fossil" isotopic signatures, we also trace the fate of radioactive materials in the environment. Surprisingly, scientists have found that, although most of the volatile fission products such as cadmium and cesium have been lost, uranium and thorium have essentially remained undisturbed. Is this because of the special geological conditions of the Oklo mine? Or does it have something to do with the very low power generation rate? There are a number of issues to be resolved for potential future applications, but these field observations could offer crucial information on designing nuclear waste disposal.

CURSE AND BLESSINGS OF POPULATION GROWTH

Many, if not all, current environmental issues, ranging from the depletion of natural resources to global climate change, can be attributed to population growth in a finite-sized space. Earthquake hazards are no exception. The death toll from earthquakes has been steadily rising over the centuries, and this is primarily because of increased population density, not because of more frequent earthquakes.[10] A highly populated city usually demands complicated infrastructures, which further aggravate its vulnerability to natural disasters. For example, a recent mega-earthquake that hit northern Japan, the 2011 Tohoku earthquake, resulted in one of the worst cases of combined natural and man-made disasters. The number of deaths and missing people caused by the earthquake and tsunami alone is considerable, exceeding 23 000 in total, but because of the subsequent disaster at the Fukushima Daiichi nuclear power plant, substantially more people have been affected by nuclear radiation, and radioactive pollution is expected to delay the recovery of the earthquake-damaged areas for many years to come.

Global climate change, energy crisis, diminishing fresh water, environmental pollution – when we list these impending issues, we have to wonder whether we can envision a bright future at all. The world is getting more and more crowded, and survival is becoming more challenging. We human beings have grown to the point at which we need to consider the consequences of our every interaction with the environment. This is happening for the first time in our history, and we need to walk untrodden paths. Population growth, however, has a bright side as well. We have more brains to tackle these difficult issues; solving conundrums one by one with creative minds is actually what has made our civilization possible. The Earth system is an extremely complex system, operating on a vast range of spatial and temporal scales, and it is not an object one can reproduce in a laboratory or in a numerical simulation. As we have seen throughout this book, however, the Earth holds innumerable traces or "fossils" of past events in a variety of guises, and every one of them, if successfully decoded, is a key to an improved understanding of the Earth system.

References

Chapter 1

1 C. Jaupart, S. Labrosse and J.-C. Mareschal, Temperatures, heat and energy in the mantle of the Earth. In *Treatise on Geophysics*, vol. 1, ed. G. Schubert. (New York: Elsevier, 2007), pp. 253–303.

2 J. S. Lewis, Consequences of the presence of sulfur in the core of the Earth. *Earth and Planetary Science Letters*, **11** (1971), 130–134.

3 M. S. T. Bukowinski, The effect of pressure on the physics and chemistry of potassium. *Geophysical Research Letters*, **3** (1976), 491–503.

4 M. T. Murrell and D. S. Burnett, Partitioning of K, U, and Th between sulfide and silicate liquids: Implications for radioactive heating of planetary cores. *Journal of Geophysical Research*, **91** (1986), 8126–8136.

5 V. Rama Murthy, W. van Westrenen and Y. Fei, Experimental evidence that potassium is a substantial radioactive heat source in planetary cores. *Nature*, **423** (2003), 163–165.

6 J. C. Lassiter, Constraints on the coupled thermal evolution of the Earth's core and mantle, the age of the inner core, and the origin of the $^{186}Os/^{188}Os$ "core signal" in plume-derived lavas. *Earth and Planetary Science Letters*, **250** (2006), 306–317.

7 C. Herzberg, K. Condie and J. Korenaga, Thermal history of the Earth and its petrological expression. *Earth and Planetary Science Letters*, **292** (2010), 79–88.

8 G. F. Davies, Thermal histories of convective Earth models and constraints on radiogenic heat production in the Earth. *Journal of Geophysical Research*, **85** (1980), 2517–2530.

9 G. Schubert, D. Stevenson and P. Cassen, Whole planet cooling and the radiogenic heat source contents of the Earth and Moon. *Journal of Geophysical Research*, **85** (1980), 2531–2538.

10 G. Schubert, D. L. Turcotte and P. Olson, *Mantle Convection in the Earth and Planets*. (New York: Cambridge University Press, 2001).

11 J. Korenaga, Urey ratio and the structure and evolution of Earth's mantle. *Reviews of Geophysics*, **46** (2008), RG2007, doi:10.1029/2007RG000241.

Chapter 2

1 J. H. Reynolds, Xenology. *Journal of Geophysical Research*, **68** (1963), 2939–2956.

2 T. Lee, D. A. Papanastassiou and G. J. Wasserburg, Demonstration of ^{26}Mg excess in Allende and evidence for ^{26}Al. *Geophysical Research Letters*, **3** (1976), 109–112.

3 A. Johansen, J. S. Oishi, M. -M. Mac Low *et al.*, Rapid planetesimal formation in turbulent circumstellar disks. *Nature*, **448** (2007), 1022–1025.

4 J. E. Chambers, Making more terrestrial planets. *Icarus*, **152** (2001), 205–224.

5 E. Kokubo and S. Ida, Oligarchic growth of protoplanets. *Icarus*, **131** (1998), 171–178.

6 F. G. Houtermans, Die Isotenhaufigkeit im natürlichen Blei und das Alter des Urans. *Naturwissenschaften*, **33** (1946), 185–186.

7 A. Holmes, An estimate of the age of the Earth. *Nature*, **57** (1946), 680–684.

8 C. C. Patterson, Age of meteorites and the Earth. *Geochimica et Cosmochimica Acta*, **10** (1956), 230–237.

9 E. Anders and N. Grevesse, Abundances of the elements: Meteoritic and solar. *Geochimica et Cosmochimica Acta*, **53** (1989), 197–214.

10 E. M. Burbidge, G. R. Burbidge, W. A. Fowler and F. Hoyle, Synthesis of the elements in stars. *Reviews of Modern Physics*, **29** (1957), 547–650.

11 D. D. Clayton, *Principles of Stellar Evolution and Nucleosynthesis*. (New York: McGraw-Hill, 1968).

12 W. S. Broecker, *How to Build a Habitable Planet*. (New York: Columbia University, 1998).

13 W. A. Fowler and F. Hoyle, Nuclear cosmochronology. *Annals of Physics*, **10** (1960), 280–302.

14 A. G. W. Cameron, The formation of the Sun and planets. *Icarus*, **1** (1962), 13–69.

15 J. J. Cowan, F. K. Thielmann and J. W. Truran, Radioactive dating of the elements. *Annual Reviews of Astronomy and Astrophysics*, **29** (1991), 447–497.

16 V. S. Safronov, *Evolution of the Protoplanetary Cloud and Formation of Earth and the Planets*. (NASA Tech. Transl. F-677, Jerusalem, 1972); translation of *Evoliutsiia Doplanetnogo Oblaka* (Nauka: Moscow, 1969).

17 T. Kusaka, T. Nakano and C. Hayashi, Growth of solid particles in the primordial solar nebula. *Progress of Theoretical Physics*, **44** (1970), 1580–1595.

18 C. Hayashi, K. Nakazawa and I. Adachi, Long-term behavior of planetesimals and the formation of the planets. *Publications of the Astronomical Society of Japan*, **29** (1977), 163–196.

19 P. Goldreich and W. R. Ward, The formation of planetesimals. *Astrophysical Journal*, **183** (1973), 1051–1062.

20 G. W. Wetherill, Formation of the terrestrial planets. *Annual Review of Astronomy and Astrophysics*, **18** (1980), 77–113.

21 G. W. Wetherill, Formation of the Earth. *Annual Review of Earth and Planetary Sciences*, **18** (1990), 205–256.

22 G. W. Wetherill, Provenance of the terrestrial planets. *Geochimica et Cosmochimica Acta*, **58** (1994), 4513–4520.

23 R. M. Canup, Dynamics of lunar formation. *Annual Review of Astronomy and Astrophysics*, **42** (2004), 441–475.

24 D. J. Stevenson, Origin of the Moon – The collision hypothesis. *Annual Review of Earth and Planetary Sciences*, **15** (1987), 271–315.

25 A. E. Ringwood, *Composition and Petrology of the Earth's Mantle*. (New York: McGraw-Hill, 1975).

26 S. R. Hart and A. Zindler, In search of a bulk-Earth composition. *Chemical Geology*, **57** (1986), 247–267.

27 C. J. Allègre, J. -P. Poirier, E. Humler and A. Hofmann, The chemical composition of the Earth. *Earth and Planetary Science Letters*, **96** (1986), 61–88.

28 W. F. McDonough and S.-S. Sun, The composition of the Earth, *Chemical Geology*, **120** (1995), 223–253.

29 T. Lyubetskaya and J. Korenaga, Chemical composition of Earth's primitive mantle and its variance: 1. Method and results. *Journal of Geophysical Research*, **112** (2007), B03211, doi:10.1029/2005JB004223.

30 L. B. Khriplovich, Eventful life of Fritz Houtermans. *Physics Today*, **45**:7 (1992), 29–37.

31 R. L. Rudnick and S. Gao, Composition of the continental crust. In *Treatise on Geochemistry*, vol. 3, ed. H. D. Holland and K. K. Turekian. (New York: Elsevier, 2003), pp. 1–64.

32 R. N. Thompson, Phase-equilibria constraints on the genesis and magmatic evolution of oceanic basalts. *Earth-Science Reviews*, **24** (1987), 161–210.

33 R. J. Kinzler and T. L. Grove, Primary magmas of mid-ocean ridge basalts. 2: Applications. *Journal of Geophysical Research*, **97** (1992), 6907–6926.

34 W. F. McDonough, Composition model for the Earth's core. In *Treatise on Geochemistry*, vol. 2, ed. H. D. Holland and K. K. Turekian. (New York: Elsevier, 2003), pp. 547–568.

35 M. Aspland, N. Grevesse, A. J. Sauval and P. Scott, The chemical composition of the Sun. *Annual Reviews of Astronomy and Astrophysics*, **47** (2009), pp. 481–552.

Chapter 3

1 Q. Yin, S. B. Jacobsen, K. Yamashita *et al.*, A short timescale for terrestrial planet formation from Hf–W chronometry of meteorites. *Nature*, **418** (2002), 949–952.

2 T. Kleine, C. Münker, K. Mezger and H. Palme, Rapid accretion and early core formation on asteroids and the terrestrial planets from Hf–W chronometry. *Nature*, **418** (2002), 952–955.

3 R. Schoenberg, B. S. Kamber, K. D. Collerson and O. Euster, New W-isotope evidence for rapid terrestrial accretion and very early core formation. *Geochimica et Cosmochimica Acta*, **66** (2002), 3151–3160.

4 D. J. Stevenson, Planetary magnetic fields: Achievements and prospects. *Space Science Reviews*, **152** (2010), 651–664.

5 B. A. Buffett, The thermal state of Earth's core. *Science*, **299** (2003), 1675–1676.

6 E. C. Bullard, The magnetic field within the earth. *Proceedings of the Royal Society of London A*, **197** (1949), 433–453.

7 P. H. Roberts and G. A. Glatzmaier, Geodynamo theory and simulations. *Reviews of Modern Physics*, **72** (2000), 1081–1123.

8 A. E. Ringwood, *Origin of the Earth and Moon*. (New York: Springer-Verlag, 1979).

9 H. C. Urey, *The Planets: Their Origin and Development*. (New Haven, CT: Yale University Press, 1952).

10 F. Birch, Speculations on the Earth's thermal history. *Geological Society of America Bulletin*, **76** (1965), 133–154.

11 A. Eucken, Physikalisch-chemische Betrachtungen über die früheste Entwicklungsgeschichte der Erde. *Nachrichten Akademischen Wissenschaften Göttingen*, Math.-phys. Klasse Heft **1** (1944), 1–25.

12 K. K. Turekian and S. P. Clark Jr, Inhomogeneous accumulation of the Earth from the primitive solar nebula. *Earth and Planetary Science Letters*, **6** (1969), 346–348.

13 E. Anders, Meteorites and the early solar system. *Annual Reviews of Astronomy and Astrophysics*, **9** (1971), 1–34.

14 H. Wänke, Constitution of terrestrial planets, *Philosophical Transactions of the Royal Society of London A*, **303** (1981), 287–302.

15 L. Grossman, Condensation in the primitive solar nebula. *Geochimica et Cosmochimica Acta*, **36** (1972), 597–619.

16 A. M. Davis and F. M. Richter, Condensation and evaporation of solar system materials. In *Treatise on Geochemistry*, vol. 1, ed. H. D. Holland, and K. K. Turekian. (New York: Elsevier, 2003), pp. 403–430.

17 M. J. Drake and K. Righter, Determining the composition of the Earth. *Nature*, **416** (2002), 39–44.

18 B. J. Wood, M. J. Walter and J. Wade, Accretion of the Earth and segregation of its core. *Nature*, **441** (2006), 825–833.

19. D. J. Stevenson, A planetary perspective on the deep Earth. *Nature*, **451** (2008), 261–265.

Chapter 4

1 J. R. de Laeter, Mass spectrometry and geochronology. *Mass Spectrometry Reviews*, **17** (1998) 97–125.

2 B. Bruhnes, Recherches sur la direction d'aimantation des roches volcaniques. *Journal de Physique*, **5** (1906), 705–724.

3 M. Matuyama, On the direction of magnetisation of basalt in Japan, Tyosen and Manchuria. *Proceedings of the Imperial Academy of Japan*, **5** (1929), 203–205.

4 T. Lee, D. A. Papanastassiou and G. J. Wasserburg, Demonstration of ^{26}Mg excess in Allende and evidence for ^{26}Al. *Geophysical Research Letters*, **3** (1976), 109–112.

5 S. C. Cande and D. V. Kent, A new geomagnetic polarity time scale for the Late Cretaceous and Cenozoic. *Journal of Geophysical Research*, **97** (1992), 13917–13951.

6 G. Faure, *Principles of Isotope Geology*, 2nd edn. (New York: Wiley, 1986).

7 A. P. Dickin, *Radiogenic Isotope Geology*, 2nd edn. (New York: Cambridge University Press, 2005).

8 E. Rutherford and F. Soddy, The cause and nature of radioactivity, part I. *Philosophical Magazine Series* **6**, 4 (1902), 370–396.

9 R. Reeves, *A Force of Nature: The Frontier Genius of Ernest Rutherford.* (New York: W. W. Norton, 2008).

10 W. Glen, *The Road to Jaramillo: Critical Years of the Revolution in Earth Science.* (Palo Alto, CA: Stanford University Press, 1982).

11 K. D. McKeegan and A. M. Davis, Early solar system chronology. In *Treatise on Geochemistry*, vol. 1, ed. H. D. Holland and K. K. Turekian. (New York: Elsevier, 2003), pp. 431–460.

12 A. N. Halliday, The origin and earliest history of the Earth. In *Treatise on Geochemistry*, vol. 1, ed. H. D. Holland and K. K. Turekian. (New York: Elsevier, 2003), pp. 509–557.

13 E. R. D. Scott, Chondrites and the protoplanetary disk. *Annual Reviews of Earth and Planetary Sciences*, **35** (2007), 577–620.

Chapter 5

1 J. R. Heirtzler and T. H. van Andel, Project FAMOUS: Its origin, programs, and setting. *Geological Society of America Bulletin*, **88** (1977), 481–487.

2 A. E. Maxwell, R. P. von Herzon, J. E. Andrews *et al.*, *Initial Reports of the Deep Sea Drilling Project*, vol. III. (Washington DC: US Government Printing Office, 1970).

3 N. Oreskes, *Plate Tectonics: An Insider's History of the Modern Theory of the Earth.* (Boulder, CO: Westview Press, 2001).

4 K. M. Creer, A reconstruction of the continents for the Upper Palaeozoic from palaeomagnetic data. *Nature*, **230** (1964), 1115–1120.

5 R. D. Müller, W. R. Roest, J.-Y. Royer, L. M. Gahagan and J. G. Sclater, Digital isochrons of the world's ocean floor. *Journal of Geophysical Research*, **102** (1997), 3211–3214.

6 A. E. Gripp and R. G. Gordon, Current plate velocities relative to the hotspots incorporating the NUVEL-1 global plate motion model. *Geophysical Research Letters*, **17** (1990), 1109–1112.

7 A. Wegener, *Die Entstehung der Kontinente und Ozeane*. (Braunschweig: Friedrich Vieweg, 1915).

8 A. Wegener, *The Origin of Continents and Oceans*. (New York: Dover, 1966).

9 K. M. Creer, E. Irving and S. K. Runcorn, Geophysical interpretation of palaeo-magnetic directions from Great Britain. *Philosophical Transactions of the Royal Society of London A*, **250** (1957), 144–156.

10 S. K. Runcorn, Palaeomagnetic comparisons between Europe and North America. *Philosophical Transactions of the Royal Society of London A*, **258** (1965), 1–11.

11 H. H. Hess, History of ocean basins. In *Petrologic Studies: A Volume to Honor of A. F. Buddington*, ed. A. E. J. Engel, H. L. James and B. F. Leonard. (Boulder, CO: Geological Society of America, 1962), pp. 599–620.

12 F. J. Vine and D. H. Matthews, Magnetic anomalies over oceanic ridges. *Nature*, **199** (1963), 947–949.

13 A. Cox and R. B. Hart, *Plate Tectonics: How It Works*. (Palo Alto, CA: Blackwell, 1986).

14 C. M. R. Fowler, *The Solid Earth: An Introduction to Global Geophysics*, 2nd edn. (New York: Cambridge University Press, 2004).

15 P. Kearey, K. A. Klepeis and F. J. Vine, *Global Tectonics*, 3rd edn. (Oxford, UK: Wiley-Blackwell, 2009).

16 J. M. Bird and B. Isacks, *Plate Tectonics: Selected Papers from Publications of the American Geophysical Union*, 2nd edn. (Washington DC: American Geophysical Union, 1980).

17 E. Orowan, Zur Kristallplastizität III: Über die Mechanismus des Gleitvorganges. *Zeitschrift für Physik*, **89** (1934), 634–659.

18 G. I. Taylor, The mechanism of plastic deformation of crystals. Part I. Theoretical. *Proceedings of the Royal Society of London A*, **145** (1934), 362–387.

19 C. Herring, Diffusional viscosity of a polycrystalline solid. *Journal of Applied Physics*, **21** (1950), 437–445.

20 N. A. Haskell, The motion of a viscous fluid under a surface load. *Physics*, **6** (1935), 265–269.

21 R. A. Daly, *Strength and Structure of the Earth*. (New York: Prentice-Hall, 1940).

22 R. B. Gordon, Diffusion creep in the Earth's mantle. *Journal of Geophysical Research*, **70** (1965), 2413–2418.

Chapter 6

1 S. A. Bowring and I. S. Williams, Priscoan (4.00–4.03 Ga) orthogneisses from north-western Canada. *Contributions to Mineralogy and Petrology*, **134** (1999), 3–16.

2 S. A. Wilde, J. W. Valley, W. H. Peck and C. M. Graham, Evidence from detrital zircons for the existence of continental crust and oceans on the Earth 4.4 Gyr ago. *Nature*, **409** (2001), 175–178.

3 V. C. Bennett, Compositional evolution of the mantle. In *Treatise on Geochemistry*, vol. 2, ed. H. D. Holland and K. K. Turekian. (New York: Elsevier, 2003), pp. 493–519.

4 J. A. Crisp, Rates of magma emplacement and volcanic output. *Journal of Volcanology and Geothermal Research*, **20** (1984), 177–211.

5 A. E. Ringwood and A. Major, Synthesis of Mg_2SiO_4–Fe_2SiO_4 spinel solid solutions. *Earth and Planetary Science Letters*, **1** (1966), 241–245.

6 C. B. Agee, Phase transformations and seismic structure in the upper mantle and transition zone. *Review of Mineralogy*, **37** (1998), 165–203.

7 G. Ito and J. J. Mahoney, Flow and melting of a heterogeneous mantle: 2. Implications for a chemically nonlayered mantle. *Earth and Planetary Science Letters*, **230** (2005), 47–63.

8 J. Korenaga, Archean geodynamics and the thermal evolution of Earth. In *Archean Geodynamics and Environments*, ed. K. Benn, J.-C. Mareschal and K. Condie (Washington DC: American Geophysical Union, 2006), pp. 7–32.

9 J. Korenaga, Urey ratio and the structure and evolution of Earth's mantle. *Reviews of Geophysics*, **46**, RG2007, doi:10.1029/2007RG000241.

10 A. M. Dziewonski and D. L. Anderson, Preliminary reference Earth model. *Physics of the Earth and Planetary Interiors*, **25** (1981), 297–356.

11 D. Porcelli and C. J. Ballentine, Models for distribution of terrestrial noble gases and evolution of the atmosphere. *Reviews in Mineralogy and Geochemistry*, **47** (2002), 411–480.

12 S. P. Grand, R. D. van der Hilst and S. Widiyantoro, Global seismic tomography: A snapshot of convection in the Earth. *GSA Today*, 7:4 (1997), 1–6.

13 A. R. McBirney, *Igneous Petrology*, 3rd edn. (Sudbury, MA: Jones and Bartlett, 2007).

14 M. Wilson, *Igneous Petrogenesis*. (Boston: Unwin Hyman, 1989).

15 S. R. Taylor and S. M. McLennan, The geochemical evolution of the continental crust. *Review of Geophysics*, **33** (1995), 241–265.

16 P. B. Kelemen, K. Hanghoj and A. R. Greene, One view of the geochemistry of subduction-related magmatic arcs, with an emphasis on primitive andesite and lower crust. In *Treatise on Geochemistry*, vol. 3, ed. H. D. Holland and K. K. Turekian. (New York: Elsevier, 2003), pp. 593–659.

17 C. J. Hawkesworth and A. I. S. Kemp, Evolution of the continental crust. *Nature*, **443** (2006), 811–817.

18 T. M. Harrison, The Hadean crust: Evidence from >4 Ga zircon. *Annual Reviews of Earth and Planetary Sciences*, **37** (2009), 479–505.

19 S. B. Jacobsen and G. J. Wasserburg, The mean age of mantle and crustal reservoirs. *Journal of Geophysical Research*, **84** (1979), 7411–7427.

20 C. J. Allègre, A. Hofmann and K. O'Nions, The argon constraints on mantle structure. *Geophysical Research Letters*, **23** (1996), 3555–3557.

21 F. Albarède and R. D. van der Hilst, Zoned mantle convection. *Philosophical Transactions of the Royal Society of London A*, **360** (2002), 2569–2592.

22 P. M. Shearer, *Introduction to Seismology*, 2nd edn. (New York: Cambridge University Press, 2009).

23 T. Lay and T. C. Wallace, *Modern Global Seismology*. (San Diego, CA: Academic, 1995).

Chapter 7

1 H. Brown, Rare gases and formation of the Earth's atmosphere. In *The Atmospheres of the Earth and Planets; papers presented at the Fiftieth Anniversary Symposium of the Yerkes Observatory, September, 1947*, ed. G. P. Kuiper. (Chicago, IL: University of Chicago Press, 1949), pp. 258–266.

2 J. W. Chamberlain and D. M. Hunten, *Theory of Planetary Atmospheres: An Introduction to Their Physics and Chemistry*, 2nd edn. (Orlando, FL: Academic Press, 1987).

3 C. F. von Weizsäcker, Über die Möglichkeit eines dualen β-Zerfalls von Kalium, *Physikalische Zeitschrift*, **38** (1937), 623–635.

4 L. T. Aldrich and A. O. Nier, Argon 40 in potassium minerals. *Physical Review*, **74** (1948), 876–877.

5 W. W. Rubey, Geologic history of sea water. *Geological Society of America Bulletin*, **62** (1951), 1111–1148.

6 S. Moorbath, J. H. Allaart, D. Bridgwater and V. R. McGregor, Rb–Sr ages of early Archaean supracrustal rocks and Amîtsoq gneisses at Isua. *Nature*, **270** (1977), 43–45.

7 J. Kunz, T. Staudacher and C. J. Allègre, Plutonium-fission xenon found in Earth's mantle. *Science*, **280** (1998), 877–880.

8 N. Takaoka and M. Ozima, Rare gas isotopic compositions in diamonds. *Nature*, **271** (1978), 45–46.

9 S. L. Miller, A production of amino acids under possible primitive Earth conditions. *Science*, **117** (1953), 528–529.

10 H. Sigurdsson (ed.), *Encyclopedia of Volcanoes*. (San Diego, CA: Academic Press, 2000).

11 G. L. Hashimoto, Y. Abe and S. Sugita, The chemical composition of the early terrestrial atmosphere: Formation of a reducing atmosphere from CI-like material. *Journal of Geophysical Research*, **112** (2007), E05010, doi:10.1029/2006JE002844.

12 P. E. Cloud, Atmospheric and hydrospheric evolution on the primitive Earth. *Science*, **160** (1968), 729–736.

13 J. C. G. Walker, The oxygen cycle. In *The Natural Environment and the Biogeochemical Cycles*, ed. O. Hutzinger. (Berlin: Springer, 1980), pp. 87–104.

14 M. Schidlowski, R. Eichmann and C. E. Junge, Precambrian sedimentary carbonates: carbon and oxygen isotope geochemistry and implications for the terrestrial oxygen budget. *Precambrian Research*, **2** (1975), 1–69.

15 R. Buick, When did oxygenic photosynthesis evolve? *Philosophical Transactions of the Royal Society of London B*, **363** (2008), 2731–2743.

16 J. Farquhar, H. Bao and M. Thiemens, Atmospheric influence of Earth's earliest sulfur cycle, *Science*, **289** (2000), 756–758.

17 M. Ozima and F. A. Podosek, *Noble Gas Geochemistry*, 2nd edn. (New York: Cambridge University Press, 2002).

18 Y. Hamano and M. Ozima, Earth-atmosphere evolution model based on Ar isotopic data. In *Terrestrial Rare Gases*, ed. E. C. Alexander Jr and M. Ozima. (Tokyo: Japan Scientific Societies Press, 1978), pp. 155–171.

19 C. J. Allègre, T. Staudacher and P. Sarda, Rare gas systematics: formation of the atmosphere, evolution and structure of the Earth's mantle. *Earth and Planetary Science Letters*, **81** (1987), 127–150.

20 E. B. Watson, J. B. Thomas and D. J. Cherniak, [40]Ar retention in the terrestrial planets. *Nature*, **449** (2007), 299–304.

21 C. F. Chyba and G. D. McDonald, The origin of life in the solar system: Current issues. *Annual Reviews of Earth and Planetary Sciences*, **23** (1995), 215–249.

22 D. W. Deamer, The first living systems: a bioenergetic perspective. *Microbiology and Molecular Biology Reviews*, **61** (1997), 239–261.

23 J. Jortner, Conditions for the emergence of life on the early Earth: summary and reflections. *Philosophical Transactions of the Royal Society of London B*, **361** (2000), 1877–1891.

24 C. Wills and J. L. Bada, *The Spark of Life, Darwin and the Primeval Soup.* (Cambridge, MA: Perseus, 2000).

25 F. J. Dyson, *Origins of Life.* (New York: Cambridge University Press, 1999).

Chapter 8

1 H. C. Urey, The thermodynamic properties of isotopic substances. *Journal of the Chemical Society* (1947), 562–581.

2 S. Epstein, R. Buchsbaum, H. A. Lowenstam and H. C. Urey, Revised carbonate-water isotopic temperature scale. *Geological Society of America Bulletin,* **64** (1953), 1315–1325.

3 M. H. Thiemens and J. E. Heidenreich III, The mass-independent fractionation of oxygen: a novel isotope effect and its possible cosmochemical implications. *Science,* **219** (1983), 1073–1075.

4 M. H. Thiemens, History and applications of mass-independent isotope effects. *Annual Reviews of Earth and Planetary Sciences,* **34** (2006), 217–262.

5 B. A. Bergquist and J. D. Blum, The odds and evens of mercury isotopes: applications of mass-dependent and mass-independent isotope fractionation. *Elements,* **5** (2009), 353–357.

6 E. Zinner, Stellar nucleosynthesis and the isotopic composition of presolar grains from primitive meteorites. *Annual Review of Earth and Planetary Sciences,* **26** (1998), 147–188.

7 J. R. Petit, J. Jouzel, D. Raynaud *et al.*, Climate and atmospheric history of the past 420,000 years from the Vostok ice core, Antarctica. *Nature* **399** (1999), 429–436.

8 T. Bernatowicz and R. M. Walker, Ancient stardust in the laboratory. *Physics Today,* **50**:12 (1997), 26–32.

9 K. McGuffie and A. Henderson-Sellers, *A Climate Modelling Primer,* 3rd edn. (Hoboken, NJ: Wiley, 2005).

10 J. Marshall and R. A. Plumb, *Atmosphere, Ocean, and Climate Dynamics: An Introductory Text.* (Boston: Elsevier Academic Press, 2007).

11 R. N. Clayton, L. Grossman and T. K. Mayeda, A component of primitive nuclear composition in carbonaceous meteorites. *Science,* **182** (1973), 485–488.

Chapter 9

1 R. F. Butler, *Paleomagnetism: Magnetic Domains to Geologic Terranes.* (Boston: Blackwell, 1992).

2 L. Néel, Théorie du traînage magnétique des ferromagnétiques en grains fins avec applications aux terres cuites. *Annales de Géophysique,* **5** (1949), 99–136.

3 G. A. Glatzmaier, Geodynamo simulations – how realistic are they? *Annual Reviews of Earth and Planetary Sciences,* **30** (2002), 237–257.

4 J. Korenaga, Urey ratio and the structure and evolution of Earth's mantle. *Reviews of Geophysics*, **46** (2008), RG2007, doi:10.1029/2007RG000241.

5 C.J. Hale and D.J. Dunlop, Evidence for an Early Archean geomagnetic field: A paleomagnetic study of the Komati Formation, Barberton Greenstone Belt, South Africa. *Geophysical Research Letters*, **11** (1984), 97–100.

6 J.A. Tarduno, R.D. Cottrell, M.K. Watkeys and D. Bauch, Geomagnetic field strength 3.2 billion years ago recorded by single silicate crystals. *Nature*, **446** (2007), 657–660.

7 R.J. Donnelly and M. Ozima, Hydromagnetic stability of flow between rotating cylinders. *Physical Review Letters*, **4** (1960), 497–498.

8 R.J. Donnelly and M. Ozima, Experiments on the stability of flow between rotating cylinders in the presence of a magnetic field. *Proceedings of the Royal Society of London A*, **266** (1962), 272–286.

9 S. Chandrasekhar, *An Introduction to the Study of Stellar Structure*. (New York: Dover, 1958).

10 S. Chandrasekhar, *Radiative Transfer*. (New York: Dover, 1960).

11 S. Chandrasekhar, *Hydrodynamic and Hydromagnetic Stability*. (New York: Dover, 1981).

Chapter 10

1 D.J. Stevenson, Origin of the Moon – The collision hypothesis. *Annual Review of Earth and Planetary Sciences*, **15** (1987), 271–315.

2 R.M. Canup, Dynamics of lunar formation. *Annual Review of Astronomy and Astrophysics*, **42** (2004), 441–475.

3 G.J.F. MacDonald, Tidal friction. *Reviews of Geophysics*, **2** (1964), 467–541.

4 S.J. Peale, Origin and evolution of the natural satellites. *Annual Reviews of Astronomy and Astrophysics*, **37** (1999), 533–602.

5 A. Barnes, Acceleration of the solar wind. *Reviews of Geophysics*, **30** (1992), 43–55.

6 G. Heiken, Petrology of lunar soils. *Reviews of Geophysics and Space Physics*, **13** (1975), 567–587.

7 B.D. Shizgal and G.G. Arkos, Nonthermal escape of the atmospheres of Venus, Earth, and Mars. *Reviews of Geophysics*, **34** (1996), 483–505.

8 K. Seki, R.C. Elphic, M. Hirahara, T. Terasawa and T. Mukai, On atmospheric loss of oxygen ions from Earth through magnetospheric processes. *Science*, **291** (2001), 1939–1941.

9 K. Hashizume and M. Chaussidon, A non-terrestrial ^{16}O-rich isotopic composition for the protosolar nebula. *Nature*, **434** (2005), 619–622.

10 T. R. Ireland, P. Holden, M. D. Norman and J. Clarke, Isotopic enhancements of ^{17}O and ^{18}O from solar wind particles in the lunar regolith. *Nature*, **440** (2006), 775–778.

11 K. Hashizume, M. Chaussidon, B. Marty and F. Robert, Solar wind record on the moon: deciphering presolar from planetary nitrogen. *Science*, **290** (2000), 1142–1145.

12 M. Abe and M. Ooe, Tidal history of the Earth–Moon dynamical system before Cambrian age. *Journal of the Geodetic Society of Japan*, **47** (2001), 514–520.

13 M. Ozima, K. Seki, N. Terada *et al.* Terrestrial nitrogen and noble gases in lunar soils. *Nature*, **436** (2005), 655–659.

Chapter 11

1 T. R. Malthus, *An Essay on the Principle of Population, as it affects the Future Improvement of Society: with remarks on the Speculations of Mr. Godwin, M. Condorset, and other writers.* (London: Printed for J. Johnson, in St. Paul's Church-yard, 1798).

2 R. A. Berner, *The Phanerozoic Carbon Cycle: CO_2 and O_2.* (New York: Oxford University Press, 2004).

3 D. Archer, M. Eby, V. Brovkin *et al.* Atmospheric lifetime of fossil fuel carbon dioxide. *Annual Reviews of Earth and Planetary Sciences*, **37** (2009), 117–134.

4 OECD World Nuclear Agency and International Atomic Energy Agency, *Uranium Resources 2003: Resources, Production and Demand.* (Paris: OECD, 2004).

5 J. C. S. Long and R. C. Ewing, Yucca Mountain: Earth-science issues at a geologic repository for high-level nuclear waste. *Annual Reviews of Earth and Planetary Sciences*, **32** (2004), 363–401.

6 R. Bodu, H. Bouzigues, N. Morin and J. P. Pfiffelmann, Sur l'existence d'anomalies isotopiques rencontrés dans l'uranium du Gabon. *Comptes Rendus de l'Académie des Sciences de Paris*, **275** (1972), 1731–1734.

7 G. W. Wetherill and M. G. Inghram, Spontaneous fission in uranium and thorium. In *Proceedings of the Conference on Nuclear Processes in Geological Settings.* (Washington, D.C.: National Research Council, 1953), pp. 30–32.

8 P. K. Kuroda, On the nuclear physical stability of the uranium minerals. *Journal of Chemical Physics*, **25** (1956), 781–782.

9 F. Gauthier-Lafaye, P. Holliger and P.-L. Blanc, Natural fission reactors in the Franceville Basin, Gabon: A review of the conditions and results of a "critical event" in a geologic system. *Geochimica et Cosmochimica Acta*, **60** (1996), 4831–4852.

10 R. Bilham, The seismic future of cities, *Bulletin of Earthquake Engineering*, **7** (2009), 839–887.

Index